Fish versus Power

Fish versus Power is an environmental history of the Fraser River (British Columbia) and the attempts to dam it for power and to defend it for salmon. Amid contemporary debates over large dam development and declines in fisheries, this book offers a case study of a river basin where development decisions did not ultimately dam the river, but rather conserved its salmon. Although the case is local, its implications are global as Evenden explores the transnational forces that shaped the river, the changing knowledge and practices of science, and the role of environmental change in shaping environmental debate. The Fraser is the world's most productive salmon river; it is also a large river with enormous waterpower potential. Very few rivers in the developed world have remained undammed. On the Fraser, however, fish – not dams – triumphed, and this book seeks to explain why.

Matthew D. Evenden is an assistant professor in the Department of Geography at the University of British Columbia, where he teaches courses in environmental history, the historical geography of Canada, and world historical geography. He has published numerous articles in journals, including the *Canadian Historical Review* and *Journal of Historical Geography*.

Studies in Environment and History

Editors
Donald Worster, University of Kansas
J. R. McNeill, Georgetown University

Donald Worster *Nature's Economy: A History of Ecological Ideas*
Kenneth F. Kiple *The Caribbean Slave: A Biological History*
Alfred W. Crosby *Ecological Imperialism: The Biological Expansion of Europe, 900–1900, 2nd Edition*
Arthur F. McEvoy *The Fisherman's Problem: Ecology and Law in the California Fisheries, 1850–1980*
Robert Harms *Games Against Nature: An Eco-Cultural History of the Nunu of Equatorial Africa*
Warren Dean *Brazil and the Struggle for Rubber: A Study in Environmental History*
Samuel P. Hays *Beauty, Health, and Permanence: Environmental Politics in the United States, 1955–1985*
Donald Worster *The Ends of the Earth: Perspectives on Modern Environmental History*
Michael Williams *Americans and Their Forests: A Historical Geography*
Timothy Silver *A New Face on the Countryside: Indians, Colonists, and Slaves in the South Atlantic Forests, 1500–1800*
Theodore Steinberg *Nature Incorporated: Industrialization and the Waters of New England*
J. R. McNeill *The Mountains of the Mediterranean World: An Environmental History*
Elinor G. K. Melville *A Plague of Sheep: Environmental Consequences of the Conquest of Mexico*
Richard H. Grove *Green Imperialism: Colonial Expansion, Tropical Island Edens and the Origins of Environmentalism, 1600–1860*
Mark Elvin and Tsui'jung Liu *Sediments of Time: Environment and Society in Chinese History*
Robert B. Marks *Tigers, Rice, Silk, and Silt: Environment and Economy in Late Imperial South China*
Thomas Dunlap *Nature and the English Diaspora*
Andrew C. Isenberg *The Destruction of the Bison: An Environmental History*
Edmund Russell *War and Nature: Fighting Humans and Insects with Chemicals from World War I to Silent Spring*
Judith Shapiro *Mao's War Against Nature: Politics and the Environment in Revolutionary China*
Adam Rome *The Bulldozer in the Countryside: Suburban Sprawl and the Rise of American Environmentalism*
Nancy J. Jacobs *Environment, Power, and Injustice: A South African History*

Fish versus Power

An Environmental History of the Fraser River

Matthew D. Evenden

University of British Columbia

CAMBRIDGE
UNIVERSITY PRESS

CAMBRIDGE UNIVERSITY PRESS
Cambridge, New York, Melbourne, Madrid, Cape Town, Singapore, São Paulo

Cambridge University Press
The Edinburgh Building, Cambridge CB2 8RU, UK

Published in the United States of America by Cambridge University Press, New York

www.cambridge.org
Information on this title: www.cambridge.org/9780521830997

First published 2004
This digitally printed version 2007

A catalogue record for this publication is available from the British Library

Library of Congress Cataloguing in Publication data
Evenden, Matthew D. (Matthew Dominic), 1971–
Fish versus power : an environmental history of the Fraser River /
Matthew D. Evenden.
p.   cm. – (Studies in environment and history)
Includes bibliographical references (p.   ).
ISBN 0-521-83099-0
1. Fishery conservation – British Columbia – Fraser River – History.
2. Pacific salmon fisheries – British Columbia – Fraser River – History.
3. Fraser River (B.C.) – Environmental conditions – History.
4. Hydroelectric power plants – Environmental aspects.
I. Title.   II. Series.
SH224.B6E83   2004
333.95´616´097113 – dc22   2003059632

ISBN 978-0-521-83099-7 hardback
ISBN 978-0-521-04103-4 paperback

To my parents and Kirsty

Contents

List of Tables, Figures, Photographs, and Maps   *page* xi

List of Abbreviations   xiii

Acknowledgments   xv

Introduction   1

1. "A Rock of Disappointment"   19

2. Damming the Tributaries   53

3. Remaking Hells Gate   84

4. Pent-Up Energy   119

5. The Power of Aluminum   149

6. Fish versus Power   179

7. The Politics of Science   231

Conclusion   267

Bibliography   277

Index   299

Tables, Figures, Photographs, and Maps

Tables

1. Fraser River Sockeye Catches on the Big-Year Cycle, 1901–33   *page* 46
2. Central Station Groups, 1942   126
3. Comparative Electrical Costs and Domestic Consumption for Systems of Equal Size in BC and Ontario, 1942   127
4. The Growth of Fish–Power Fisheries Research Measured by Project Starts   245
5. Research Institutions and Cooperative Projects   246

Figures

1. BCER mainland power production (in kilowatt hours), 1905–28   65
2. "Fish Travel Modern Highway"   114
3. "The Future Fraser River"   187

Photographs

1. Hells Gate, ca. 1867   22
2. "Rockslide, Fraser River at Hells Gate"   27
3. Sockeye Salmon in Spuzzum Creek, August 1913   29
4. Clearing the Gate, 1914   33

5. Opening of the tunnel connecting Coquitlam and
   Buntzen Lakes   61
6. Dr. William Ricker at Hells Gate, 1938   85
7. Tagged sockeye salmon from the Hells Gate
   investigations   92

Maps

1. Fraser Basin and British Columbia   6
2. Hells Gate and the landslides   25
3. The Fraser Canyon, ca. 1916   26
4. Lower-basin dam projects   57
5. Flooding in the Fraser Valley, 1948   143
6. The Alcan project   171
7. The Columbia River projects   181
8. Transmission networks in BC, 1958   190
9. Fraser Basin Board power site investigations, 1958   224

Abbreviations

Alcan	Aluminum Company of Canada
BC	British Columbia
BCER	British Columbia Electric Railway Company
BCPC	British Columbia Power Commission
BPA	Bonneville Power Administration
CCF	Canadian Commonwealth Federation
CNR	Canadian Northern Railway
CPR	Canadian Pacific Railway
FRBC	Fisheries Research Board of Canada
IDA	Industrial Development Act
IPSFC	International Pacific Salmon Fisheries Commission
IJC	International Joint Commission
IWA	International Woodworkers of America
MLA	Member of the Legislative Assembly
MP	Member of Parliament
PUC	Public Utilities Commission
REA	Rural Electrification Administration
REC	Rural Electrification Committee
TVA	Tennessee Valley Authority
UBC	University of British Columbia
UFAWU	United Fishermen and Allied Workers Union
UW	University of Washington
VMAD	virgin mean annual discharge
WCPC	Western Canada Power Company

Acknowledgments

It gives me great pleasure to thank the individuals and institutions who helped to bring this book to completion. As a graduate student I received assistance from the Social Sciences and Humanities Research Council Doctoral Fellowships program, the Canadian Forest Service Graduate Supplement initiative, and York University, including the President's Scholarship and the Ramsay Cook Scholarship. At the University of British Columbia, several small Hampton grants, and funds from the dean of arts and my department, have assisted me in my research. I gratefully acknowledge permission to use materials previously published in the *Journal of Historical Geography* (Elsevier) and *BC Studies*, in Chapters 1 and 3, respectively. I also thank the Pacific Newspaper Group, the Pacific Salmon Commission, and the BC Archives for permission to republish photographs and images. I have made every reasonable effort to determine the copyright holders of the photographs and images. If any person has information on the rightful copyright holders, I would be pleased to hear from them.

This book has been written with the support of numerous teachers, colleagues, and friends. I first became interested in environmental history as an undergraduate because of the brilliant teaching and example of Colin Duncan. A wide group of historians at York inspired me in my graduate studies, sometimes from remarkably different perspectives. I thank, in particular, Christopher Armstrong, Ramsay Cook, Craig Heron, Richard Hoffmann, Richard Jarrell, Elinor Melville, and Marlene Shore. I would also like to thank my fellow students and friends Dimitry Anastakis, Sarah Elvins, and James Muir. Archivists at all of the institutions listed in the notes provided considerable help. George Brandak of the University of British Columbia Special Collections and Archives and Teri Tarita of the Pacific Salmon Commission Library deserve special thanks. At the University of British Columbia, I have benefited from the advice and support of colleagues and students in the Department of Geography. Two heads, Graeme Wynn and Mike Bovis, supported my work in various ways. Eric Leinberger prepared the maps and helped me to sort out cartographic problems. Cain Allen, Justin Barer, and John Thistle provided research assistance. Several people read this book in its first incarnation as a thesis: Christopher Armstrong, Stephen Bocking, John Chapman, Richard Jarrell, Elinor Melville, Anders Sandberg, and Marlene Shore. Christopher Armstrong, Cole Harris, Doug Harris, and Kirsty Johnston read several of the manuscript chapters. Cole Harris also introduced me to the Fraser Canyon in a new way. Cain Allen read the entire manuscript and sharpened my analysis of the Columbia River. Graeme Wynn read the whole manuscript once and several chapters a second time amidst considerable administrative responsibilities. He has been enormously encouraging. I owe a great debt to Viv Nelles, who supervised my PhD and provided excellent advice and support at every stage. I could not imagine a better supervisor. Series editors John McNeill and Donald Worster provided encouragement and good advice. The anonymous reviewers provided useful criticism. Editor Frank Smith and his assistant Eric Crahan of Cambridge University Press guided a novice through the process with skill and patience. The production and copyediting team at TechBooks proved enormously efficient and helpful. Vicki Danahy had a scrupulous eye for detail and Eleanor Umali kept the process moving and answered all of my questions. Of course, I alone am responsible for any errors of fact or interpretation that remain.

Fish versus Power brought me home to British Columbia and reminded me of the unshakable support of family. My sisters Kirstin and Maya and their husbands Chris and Steve helped out in numerous ways. My uncle, John Walters, assisted me with various copyright matters. My parents welcomed me home, traveled with me around the Fraser Basin, and listened to long oral reports over dinner. They have shown me every encouragement and have demonstrated throughout my life the importance of ideas. Kirsty Johnston has been hearing about fish and power for years in various apartments, several cities, and three provinces. The Fraser has flowed through it all. She has helped me in too many ways to list and cheered me always. I am grateful for her love and support.

I have one last thank you to make to the artist, my late grandfather, John Paton Walters, whose painting of the Fraser Canyon in 1969 has been reproduced on the cover. I found this painting at my parents' house in Vancouver in 2003. It had been stored years ago and forgotten, one piece among a small collection of BC landscapes. I was astonished to see that he had painted a place, looking out from Lytton, up the Fraser, that I had visited many times on field trips. It is a striking point at the confluence of the Thompson and Fraser and it must have impressed him when he saw it during one of his visits to the coast. He was used to painting tide mills and shorelines in Suffolk and here was the rough texture of the canyon and the stirring, overlapping colors of two rivers combined. So imagine my delight when I curled back the brown covering on this painting and discovered that the river I had thought so much about had been represented so beautifully by my grandfather.

Introduction

In the twentieth century, humans transformed the planet's rivers. On every continent, save Antarctica, they dammed, diverted, and depleted rivers. On local, regional, and continental levels, the pathways and uses of water changed, if not always in kind, then in intensity, location, and scale. Two ecologists, Mats Dynesius and Christer Nilsson, reported in the closing decade of the twentieth century that over three-quarters of the total water discharge of the 139 largest rivers in the Northern Hemisphere are "strongly or moderately affected by fragmentation of the river channels by dams and by water regulation resulting from reservoir operation, inter-basin diversion, and irrigation....These conditions indicate that many types of river ecosystems have been lost and that the populations of many riverine species have become highly fragmented."[1]

The aim was not to fragment but to create a new order. In emerging nation–states, in empires and colonies, in capitalist and communist societies, political elites applied new technologies of power to rivers and lakes. Dreams of a hydraulic order sought to correct past ills, to raise wealth, to impose control over nature and others. Floods would be stopped, rushing, wasting water would be harnessed with hydroelectric dams, and arid lands and reservoirs would be linked with irrigation systems. A new order of previously unimaginable scope was placed on rivers almost everywhere. Over the course of the twentieth century, humans increased their annual

---

1. Dynesius and Nilsson, "Fragmentation and Flow," p. 753. For Dynesius and Nilsson, "northern third of the world" refers to North America north of Mexico, Europe, and the republics of the former Soviet Union.

1

withdrawals from rivers and lakes ninefold.[2] And still the search for order, on the Yangtze, on the Mekong, and on the Great Plains, continues.

Because so many dams have been built and so many rivers transformed, the historiography of water development has understandably focused on the factors driving development rather than on those constraining it. From the role of state systems and geopolitics to the importance of monumentalism and modernism in dam design, environmental historians have sought to understand the forces that have made dam development politically possible, economically feasible, and ideologically acceptable.[3] Complementing a number of important studies focusing on fisheries, they have also catalogued a host of environmental consequences of river development, not only on fish, but also on changed flow regimes and on human settlements.[4] Relatively few national and international studies have sought to resurrect and understand protest movements against dam development.[5] In general, environmental historians of river development have emphasized the causes of development without questioning what the alternatives might have been or whether, in similar places at similar times, the outcomes were the same.

This book focuses on a river for which some of the choices made about how to impose order, to raise wealth, and to establish political power took different forms than those used elsewhere. The Fraser River, located in British Columbia, Canada, is the most productive salmon river in the world. In terms of annual discharge, it is the third largest river on the Pacific coast of North America, following the Columbia and the Yukon, and the third largest in Canada, after the Mackenzie and the St. Lawrence. Like many large North American rivers, the Fraser has inspired dreams of waterpower wealth and fisheries growth. Despite many attempts in

---

2. McNeill, *Something New Under the Sun*, pp. 120–1. Estimates of long-term changes in human uses of freshwater may be found in L'vovich and White, "Use and Transformation," pp. 235–252.
3. Blatter and Ingram, *Reflections on Water*; Goldsmith and Hildyard, *Social and Environmental Effects*; Headrick, *The Tentacles of Progress*; Hundley, *The Great Thirst*; Jackson, *Building the Ultimate Dam*; Nelles, *The Politics of Development*; Pisani, *To Reclaim a Divided West*; Tyrrell, *True Gardens of the Gods*; Worster, *Rivers of Empire*.
4. McEvoy, *The Fisherman's Problem*; Taylor, *Making Salmon*; White, *The Organic Machine*.
5. Harvey, *A Symbol of Wilderness*; McCully, *Silenced Rivers*.

the twentieth century, however, the main stem of the Fraser has never been dammed. The river has been fragmented, but far less than most. For students of the modern world and of the environment, the Fraser River provides an important counterpoint to other rivers in North America and beyond. Its many developmental similarities highlight its important differences. On the Fraser, fish have triumphed, not dams, and this study seeks to explain why.

The answer turns on the political economic and transnational forces that shaped the Fraser River in the twentieth century. Before 1945, developers dammed tributaries to assist resource extraction activities and to power urban growth. However, the limits of local markets and state intervention as well as the practical difficulties of dam development kept large dams off the Fraser. After 1945, power demand soared and flood control became a major goal of public policy. Diverse interests promoted the benefits of damming the river. In this high stage of postwar development pressure, a fisheries conservation coalition, born of a fishing industry alliance, Canada–U.S. cooperation in the international regulation of the salmon fishery, and the intervention of fisheries scientists critical of the effects of power development, held dams off the Fraser. Once Canada concluded an international treaty (1964) with the United States to dam the upper Columbia in coordination with American projects downstream, the British Columbia government proceeded with another major power project on the northern Peace River, which would meet domestic and export power demands for decades. As a result, the Fraser became insulated from development pressures. Former dam proposals could no longer be justified economically or politically. Development was delayed, then displaced to other rivers.

A second argument of this book is that the fish vs. power debate not only delayed and displaced development, but also produced various side effects on salmon, science, and society. Over the course of the twentieth century, debates about how to manage the river and the salmon bore consequences for salmon migration when landslides blocked the main stem, for salmon spawning grounds when developers dammed tributaries, and for salmon populations when restoration activities reclaimed areas for salmon rearing. Throughout the fish vs. power debate, scientists played an important role in analyzing potential development cases, providing expert advice, and studying the relevant problems. Scientific institutions were formed and strengthened in part to provide answers to the fish–power problem. The applied questions that scientists sought to answer directed their research along some avenues and not others. In the background to the development conflict lay basic questions about British Columbia's future. The fish vs. power debate broke and forged coalitions among different social groups, regional interests, industries, and state bureaucracies. Because the Fraser salmon fishery was international and because the development of the Fraser would affect other transboundary developments on the Columbia, the fish vs. power debate expanded into national and international politics, challenging and provoking older definitions of regionalism and regional self-interest. The debate over the river and its fish, in short, climbed the banks of the Fraser and affected and interacted with a range of problems and processes beyond.

"You cannot step twice in the same river," observed the pre-Socratic philosopher Heraclitus, "for other waters are continually flowing on."[6] Writing a river's history, then, seems a peculiarly elusive project, chasing a flow that cannot be constrained. Yet rivers have served as both metaphors and subjects of history for thousands of years. In Canada, rivers became routes for an unfolding nation in an earlier historiography; today they appear more frequently in our writing as sites of pollution.[7] Romantic highways have become sewers. In British Columbia, rivers flow through the contemporary environmental imagination. They carry salmon that are said to typify a region and its history. They connect people to a place and its past. Frequently British Columbians have tried to step into the same river twice, forgetting how the river has changed and, more frequently still, how they have changed with it.

Before 1900, when this book begins, modernizing forces of the world economy and the politics of colonization had already

6. Quoted in Wheelwright, *Heraclitus*, p. 29.
7. The classic statement of the Laurentian interpretation of Canadian history is found in Creighton, *The Commercial Empire*; New, "The Great River Theory."

changed the river and its peoples. To gain a long-term perspective that helps to place the development conflicts of the twentieth century in context, it is well to describe some of the natural processes that have made this river and to reflect upon the forces that turned the Fraser into a resettlement corridor.

The Fraser River drains a large basin, 250,000 km² (1 km² = 0.3861 miles²), about a quarter of British Columbia, or slightly larger than the land area of the United Kingdom (see Map 1). Since 1952, when the Nechako River was partially diverted to the coast, the drainage area has practically decreased to about 233,000 km².[8] The Fraser originates in the western slopes of the Rocky Mountains near Mount Robson and curves in a long s-shaped southwestern arc toward the delta, 1375 km away. Over the seasonal cycle, the river's mean flow, measured near the river mouth, changes dramatically from 750 m³/s in the winter to 11,500 m³/s in the summer (1 m³ = 1.308 yd³).[9] From mountainous slopes in the western Rockies, the river cuts deeply into interior plateaus and enters a wide valley near the delta. In its course, the river passes through several different ecological regions shaped by varying conditions of moisture and temperature: evergreen forests of pine, spruce, and fir on the western slopes of the Rockies, dry plateau and sagebrush country in some sections of the interior, and dense rain forests on the coast. Each of these areas provide rich and varied habitat for salmon. Five species of Pacific salmon (*Oncorhynchus*) spawn in the Fraser basin: chinook (*O. tshawytscha*), chum (*O. keta*), coho (*O. kisutch*), pink (*O. gorbuscha*), and sockeye (*O. nerka*).[10] Pacific salmon are anadromous. They spend their early life in freshwater environments, usually food-rich lakes or small streams, migrate to the ocean for the bulk of their life history, and, in their final phase, return to their natal streams to spawn and die.[11]

8. Church, "The Future of the Fraser River."
9. Dorcey, "Water"; Moore, "Hydrology and Water Supply," Vol. 2, pp. 3–18, 21–40; Northcote and Larkin, "The Fraser River."
10. I deliberately leave steelhead/rainbow trout out of this listing, following Groot and Margolis: "The scientific names for steelhead/rainbow trout (*Salmo gairdneri*) and cutthroat trout (*Salmo clarki clarki*) have recently been changed. They are now included in the genus *Oncorhynchus* and have been renamed *O. mykiss* and *O. clarki*, respectively. ... They are still known as trout rather than salmon." Groot and Margolis, *Pacific Salmon Life Histories*, p. x.
11. Ibid.

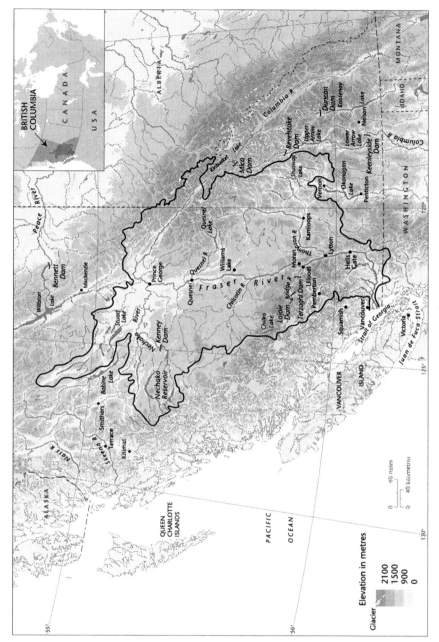

Map 1. Fraser Basin and British Columbia.

The Fraser River entered the modern world system relatively late. Native peoples had lived along the river for millennia and developed complex social worlds around the primary resource of salmon. In the lower sections of the river, salmon and a diverse marine environment provided the foundations for some of the largest native populations in Canada before contact. Coastal contacts between native peoples and European explorers and traders occurred only in the late eighteenth century – more than 200 years later than similar encounters in Eastern Canada. Smallpox, spreading north and west from the Plains in 1782, preceded direct contact in many areas of the coast and Fraser River and had devastating consequences for native groups.[12] Land-based contacts, driven by an expansive continental fur trade from the East, were not realized until the early nineteenth century when Simon Fraser descended the river that would later bear his name, mistaking it for the Columbia. Although the treacherous narrows of the Fraser Canyon ruled out the use of the river as a possible transportation corridor for the fur trade, fur-trade posts were established in the interior and near the delta. These early links connected peoples and environments of the Fraser Basin to a continental trade network and to the demands of international markets, but they did not displace native peoples, nor colonize them.

A gold rush up the Fraser River in 1858 changed all of this. In a matter of months, around 20,000 miners, mainly from California, but also from eastern North America and Asia, traveled up the river, staked claims, and developed placer mining sites. Numerous conflicts with native peoples along the river ensued. In the same year, the British created the colony of British Columbia to exert control. The Fraser Basin lay within the territory ceded by the United States to the British in the Treaty of Washington (1846). The British established a settlement system, built a wagon road along the river from Yale (the head of navigation) north to the Cariboo gold fields, surveyed town sites, and generally imposed a new system of law and order.[13] Missionaries of several denominations seeking cultural and religious change followed. As part of this

---

12. Harris, "Voices of Smallpox Around the Strait of Georgia," in *The Resettlement*, pp. 3–30.
13. Loo, *Making Law*.

resettlement process, in the second half of the nineteenth century native peoples were confined to margins of their traditional territories on reserves; unlike in many other parts of Canada, here no treaties were signed.[14] In a very short period of time, the river had been colonized, connected to a global empire, and exploited for a high-value, low-bulk commodity export.

The Fraser River, nevertheless, remained on the margins. Although small resource industries of lumbering, fishing, agriculture, and mining developed slowly, particularly in the lower sections of the river, the lack of substantial local markets and the prohibitive costs of transportion to larger centers limited opportunities. It was not until the late nineteenth century, after the completion of the Canadian Pacific Railway, a transcontinental railroad, that the barriers of distance began to diminish. From the east, the railroad followed the Thompson River, a Fraser tributary. At the confluence of the Thompson and the Fraser at Lytton, the railroad cut south, following the Fraser to the coast, before terminating north of the delta, on Burrard Inlet. At its terminus, finally reached in 1887, the railroad provided the economic foundations for a new city, Vancouver. At the coast, the railroad connected with steamships. Along its tracks, telegraph lines carried information. Over the years, a series of spur lines ran up tributary valleys that bisected the main line. In the early twentieth century, a second transcontinental line followed the river. The railroad, however, did not move the Fraser River from the margins to the center. Beyond rail lines, travel and shipment in the Fraser Basin remained difficult. But the railroad did connect the region to a new East–West axis across northern North America that promised considerable political and economic change.

The construction of the railroad had been a fundamental condition of British Columbia's entrance into Canadian Confederation in 1871. Before that time, the mainland colony had been united with Vancouver Island in 1866. By entering Confederation, British Columbia became subject to the constitution of a complex federal state. By the terms of the British North America Act, 1867, provinces and the federal government gained different

14. Harris, *Making Native Space.*

responsibilities and powers.[15] Section 109 of the Act granted provinces jurisdiction over lands and resources, and Sections 92(5) and 92(13) granted the right to manage and sell public lands and to hold jurisdiction over property and civil rights. The federal government, however, retained jurisdiction over ocean-based and anadromous fisheries, navigable and international rivers, and significant powers related to natural resources under Section 91, particularly as these related to trade and commerce, taxation policy, and, most broadly, "Peace, Order and good Government." Further, a railway belt, set aside to help finance the transcontinental railroad, remained under federal authority. The belt extended 20 miles on either side of the main line and therefore encompassed a wide section of the Fraser Basin. Administration of the railway belt remained contested for many years.[16] These divisions of authority produced various conflicts within the state system over the administration and regulation of shared or contested resources – conflicts that private interests happily exploited.[17]

The railroad provided relatively cheap, reliable transport that underwrote a new era of resource development in the Fraser Basin. The primary exports from the basin were forest products and fish. Lumbering centered on the lower river and its environs, whereas fishing occurred near the delta and focused primarily on sockeye salmon. Agricultural commodities wrested from the rich soils of the Fraser Valley reached local markets in resource towns and the expanding commercial center of Vancouver. Before the turn of the century, cattle-ranching operations moved into sections of the middle basin, along the Thompson River. Rapidly, the railroad opened new resource development opportunities in the Fraser Basin and made it into an export-oriented, staple-producing region.

At the edge of the transcontinental railroad system, Vancouver emerged as the primary Pacific metropolis of Canada, organizing the provincial resource trades and providing transportation services for a vast hinterland. Within the city, lumber mills cut timber from logs floated by river and sea and prepared them for shipment by

15. The British North America Act, 1867, *Statutes of the United Kingdom*, 30–1, Victoria, Chapter 3.
16. Cail, *Land, Man, and the Law*, pp. 111–52.
17. Hessing and Howlett, *Canadian Natural Resource*, p. 56.

rail and barge. To the south, on the river, an increasing number of canneries produced an expanding pack of sockeye salmon. Cases of tinned fish were moved by rail to consumers in the east and abroad. Although manufacturing remained limited in the city before 1900, the city grew rapidly. Vancouver's population more than doubled from a modest 13,709 in 1891 to almost 30,000 by the turn of the century. In the next decade, the city's population quadrupled. By 1931, the city and surrounding municipalities reached nearly a quarter of a million; by 1951, over half a million – or nearly one-half of the entire provincial population.[18] A city of this size and regional importance provided a considerable market for the resource economy of the province and also developed a substantial service industry and manufacturing base.

Fish versus Power is an environmental history of a contested river. To understand that contest, I build upon and seek connections among three bodies of scholarship.

First, any study of primary resource development in Canada must come to terms with the staples tradition of Canadian political economy. This tradition can be traced to the foundational economic histories of Harold Innis on Canada's early export trades of fur, cod, and minerals. Innis analyzed the "penetrative powers of the price system" across a vast northern realm and worked out the effects of staple commodity exploitation on economic and social relationships in hinterland regions.[19] He approached these problems with the assumption that economic activity was linked to and embedded in environmental and spatial contexts. Although Innis's work has been widely praised and remains important, those who have sought to emulate his approach and concerns have done so in broadly different ways. In one direction, political economists have largely abandoned an historical approach, but have given greater theoretical precision to Innis's ideas and emphasized the inherent difficulties of industrialization in a staple-driven peripheral

18. McDonald, *Making Vancouver*; Wynn, "The Rise of Vancouver," p. 69.
19. For a general introduction to Innis's work and a collection of his essays, see Drache, ed., *Staples, Markets and Cultural Change* (Kingston/Montreal: McGill-Queen's University Press, 1995).

economy.[20] In Canadian historical writing, the staples approach has received scattered treatment in the past half century; the most notable reinvigoration has appeared in the work of H. V. Nelles, who connected comparative studies of staple trades to analyses of the state and regulation.[21] In historical geography, the staples tradition attracted considerable interest in the 1970s and 1980s and emphasized the settlement patterns associated with staple trades, as well as the environmental contexts and effects of commodity exploitation.[22] My own approach analyzes the connections between what Nelles has called the "politics of development" in a federal state and the environmental histories of contested resources.

Second, a body of work in transnational environmental history has raised important questions. Although the term transnational has come to mean different things across the social sciences and humanities, in environmental history it refers to the connections made between humans and environments across national boundaries. Ian Tyrrell's recent study of environmental exchanges between Australia and California, for example, explores the complex interplay of cultural meanings and ecological processes at various spatial scales.[23] The problems and practices of a transnational approach have attracted the greatest attention in American historiography and challenged persistent notions of American exceptionalism.[24] Beyond the United States, explicit reference to transnational concerns has been less evident, although there is no doubt that world history scholarship and numerous national (including Canadian) historiographies have treated transnational problems implicitly, or as a matter of course. Although this book focuses on one river basin, prescribed political or bioregional units of analysis do not confine it. I address the local, national, and international forces that have shaped the river and have been shaped by it. By making the transnational aspects explicit, I suggest the importance of understanding seemingly local or seemingly international

20. Drache, "Celebrating Innis: The Man, The Legacy and Our Future," in ibid., pp. xiii–lix.
21. Nelles, *The Politics of Development.*
22. Harris, "Industry and the Good Life Around Idaho Peak," *Canadian Historical Review* 66(3) (1985): 315–43; Ray, *Indians in the Fur Trade*; Wynn, *Timber Colony.*
23. Tyrrell, *True Gardens of the Gods.*
24. Tyrrell, "Making Nations/Making States"; idem, "American Exceptionalism"; White, "The Nationalization of Nature."

environmental questions at different and frequently intersecting spatial scales.

Third, recent work in environmental history has drawn connections among environmental change and the ideas and practices of science.[25] Broadly, environmental historians have borrowed selectively from recent constructivist approaches to the history and sociology of science that emphasize the contextual aspects of knowledge and practice. They have extended the range of problems under analysis, however, by linking scientific ideas and practices to changing environmental contexts.[26] Joseph Taylor's environmental history of the Columbia River fisheries, for example, identifies the agency of scientific knowledge in refashioning salmon populations under various fish cultural techniques.[27] Nancy Langston's environmental history of forestry in the U.S. West, on the other hand, demonstrates the mutually constitutive roles of changing scientific ideas and management practices acting on and changing within a shifting forest regime.[28] This book builds on this previous work by linking environmental history and the history of science, while also attending to the institutional and political contexts of scientific knowledge.

This book also depends critically on a rich body of local scholarship. Perhaps the first historian of fish and power was Henry Doyle, a British Columbia canner and self-taught scholar who produced a history of the Pacific Coast fisheries in the mid-1950s in the hopes of educating the public and warding off the power threat. Although he tried to publish this history, no press would take it, chiefly because of its heterodox theories of salmon biology, but also perhaps because of its fierce denunciations of the power interest.[29] Since Doyle's polemic, the subject has passed. Because there are no dams on the Fraser, presumably, their absence requires little explanation. Historians of hydroelectricity have focused on the rivers on which development did take place.[30] Historians of British

25. White, "Environmental History."   26. Golinski, *Making Natural Knowledge.*
27. Taylor, *Making Salmon.*   28. Langston, *Forest Dreams, Forest Nightmares.*
29. UBC Special Collections and Archives (hereafter UBC), Doyle, Henry Papers, *The Rise and Decline of the Pacific Salmon Fisheries,* 2 vols. (nd, 1957?), unpublished MS.
30. For the early period, see Armstrong and Nelles, *Monopoly's Moment*; Roy, "The British Columbia Electric Railway Company"; idem, "The Fine Arts of Lobbying"; idem, "The Illumination of Victoria"; idem, "The British Columbia Electric Railway and Its Street Railway Employees"; idem, "Direct Management from Abroad." For the

Columbia's fisheries have provided a number of excellent studies on the role of business in the commercial fishery and on aspects of ethnic and native history.[31] Few have investigated how nearly the industry came to an end in the face of dam development.[32] In terms of the environmental conditions of the fishery, my attention is focused on the river. Although important studies over the past decade have provided new ideas and evidence about the effects of changing ocean conditions on salmon life history and productivity, much uncertainty remains around specific instances of environmental change, such as the effects of the Hells Gate slides.[33]

The Fraser's present and future have also shaped my questions and concerns. The Fraser River's salmon fisheries, like so many fisheries worldwide, have experienced sharp declines in recent years. In 1999, the sockeye fishery was canceled for the first time. Concurrently, British Columbians and North Americans increased their appetite for energy. As California struggles to secure energy supplies, British Columbia and other Canadian jurisdictions export surplus power and imagine ways to develop and sell more.[34] With a faltering fishery on the Fraser and increasing energy demands, can calls for power dams on the Fraser be far behind? Today, the Fraser provides a counterexample to many rivers elsewhere, but it still has the potential to follow more familiar patterns.

Fish versus Power opens with a tragedy: a series of devastating landslides in 1912–14 that dammed the river in a narrow gorge named Hells Gate. The slides cut off the migration path of the

postwar period, see Bocking, *Mighty River*; Mitchell, *WAC Bennett*; Mouat, *The Business of Power*; Swainson, *Conflict Over the Columbia*; Wedley, "The Wenner-Gren and Peace River Power Development Programs"; idem, "Infrastructure and Resources"; Williston and Keller, *Forests, Power and Policy*, especially Chapter 2, "The Two Rivers Policy."

31. Harris, *Fish, Law, and Colonialism*; Johnstone, *The Aquatic Explorers*; Meggs, *Salmon*; Muszynski, *Cheap Wage Labour*; Newell, *Tangled Webs*; idem, *The Development of the Pacific Salmon-Canning Industry*; idem, "The Politics of Food in World War II"; idem, "Dispersal and Concentration"; idem, "The Rationality of Mechanization"; Ralston, "Patterns of Trade"; Reid, "Company Mergers."

32. Writers reaching a broader popular audience have raised some of these questions, however: see Bocking, *Mighty River*; Roos, *Restoring Fraser River Salmon*.

33. For an important statement on this area of research, see Beamish and Bouillon, "Pacific salmon production trends."

34. Froschauer, *White Gold*.

largest salmon run ever recorded. The Fraser Canyon filled with the writhing red bodies of spawning salmon, visible for 10 miles beneath the obstruction. In combination with other factors, the slides precipitated the collapse of the Fraser fisheries. For thirty years salmon runs fell below a quarter of their historic levels. Native peoples, commercial fishers, fisheries scientists, and regulators adjusted to the new conditions and sought solutions. Chapter 1 examines the causes and consequences of the slides and lays the foundations for understanding subsequent attempts to restore the river.

Visions of a transformed river abounded after 1900. Power interests surveyed the river's flow, sited dams, and promoted development schemes. Small dams rose on tributaries to facilitate mining, forestry, and hydroelectric concerns. Before 1940 these projects proceeded with little regulatory oversight or constraint and had adverse effects on important spawning grounds. Main-stem projects, however, remained unrealized. Weak market demand and the focused development projects of a monopoly utility in the metropolitan regions of the province did not provide the economic conditions to support major dam projects. No government intervention in power development propelled a building program. Chapter 2 investigates the course of dam development on Fraser tributaries in the early twentieth century and explains why the main stem remained undammed.

After the decline of the Fraser sockeye runs following the Hells Gate slides, Canada and the United States negotiated the Pacific Salmon Convention (1937) to establish a catch agreement and to launch a scientific restoration program under the International Pacific Salmon Fisheries Commission. Following scientific investigations that identified blockage problems at Hells Gate, the International Pacific Salmon Fisheries Commission built fishways to overcome the difficulties. Here was the first of several instances in which the transnational nature of the resource affected local environmental management. Just as pressure built in British Columbia to proceed with major dam development on the Fraser in the late 1940s, salmon runs rebounded and Hells Gate emerged as a model of the costs of development. In Chapter 3, I examine the development of a transnational research program, the restoration of Hells Gate, and the local and international controversies in which this program became embroiled. I argue that the construction of the fishways created a significant claim to the Fraser Canyon just as political pressures mounted to dam the Fraser.

The growth of British Columbia's economy during World War II inspired new demand for electrical energy, public power development, and expansion into hinterland regions. Canadians looked enviously upon regional growth in the U.S. Pacific Northwest spurred by dam projects and associated industrial development. Calls were made to nationalize the power sector, to dam the Fraser, and to drive the region into a new industrial phase. Urban centers expanded, provincial population grew, and the resource economy prepared for a postwar development phase. In response, BC Electric launched a major building campaign and new developers entered the scene in the hopes of harnessing British Columbia's waterpower wealth. Chapter 4 suggests how the context of a world war and the mounting demands of the postwar period transformed the possibilities for development on the Fraser.

Overlapping resource demands made the Fraser River a contested site of development politics. By the late 1940s, the fish vs. power debate had begun. In the first major dispute of the postwar era, the Aluminum Company of Canada successfully obtained rights to develop the Nechako River, an upper-basin tributary. With strong provincial support, the Aluminum Company's development scheme promised to create the third largest city in the province, propel industrial development in the provincial North, and affect salmon runs minimally. The fisheries defense proved scattered and contradictory. As the Aluminum Company's project rose in the North, flooding out traplines and burial grounds of Cheslatta T'en natives and diverting the Nechako with consequences for salmon runs downstream, the development agenda appeared unstoppable. In Chapter 5, I examine the politics of conflicting resource demands and demonstrate how the aluminum industry and provincial government co-opted fisheries protest and federal opposition. Despite the loss of fisheries interests in this case, the Aluminum Company dispute provided the opportunity for the organization of a cross-industry fisheries coalition, made up of labor, capital, scientists, state officials, and politicians.

During the 1950s a host of promoters and corporations proposed to dam the river many times over. Federal officials investigated the diversion of the Columbia River into the Fraser and damming the massive combined flow. The provincial government examined flood-protection dams and power dams on tributaries to expand regional electricity supplies. Major utilities conducted development surveys and funded a "crash program" of research to solve the fish–power problem. Salmon, dam proponents argued, would have to be "retrained" or "go the way of the buffalo." The threat of unfettered river development helped to unify an otherwise fractious set of fisheries interests. Native peoples, fishers, capitalist canners, communist union leaders, scientists, American lawyers, and federal politicians joined together to criticize the damming program and ensure that alternatives received appropriate attention. Antidam forces looked to the future of atomic power or cast their eyes to the provincial interior and north in search of other possible developments on the Peace and Columbia Rivers, rather than damming the Fraser. In the same period, the United States proposed cooperative development on the upper Columbia to control river flows to benefit projects downstream. Critics of these alternatives charged that transmission costs from distant projects were unfeasible, that fish and dams could exist in harmony, and that developing the Columbia with the United States would sell Canadian interests "down the river." In Chapter 6, I analyze some of the most spectacular of these development plans, consider their popular reception, and connect the Fraser River to other national and international agendas.

Scientists played a major role in the fish vs. power debate. In attempting to influence a public debate, they changed the public profile of science and faced questions about the boundaries between lay and expert knowledge. The debate inspired an enormous investment in research to pass fish over dams, which provided the opportunity to expand the institutional basis of fisheries science in British Columbia and launch research careers. The debate also focused scientific attention on problems of freshwater habitat, on physiology and ecology, and on applied studies over basic science.

In Chapter 7 I argue that scientists changed the fish vs. power debate and that the debate changed science.

In the Conclusion, I compare the case of the Fraser with other North American and world rivers. In particular, I examine how the Fraser compares with its closest parallel case – the Columbia River – and consider how development politics differed on these two large rivers, producing divergent outcomes. I argue that the Fraser River was protected as a salmon river in spite of considerable development forces by the cooperation of local and international political interests and the possibility of alternative development sites within British Columbia.

The book ends roughly in 1960, at a time, I suggest, when the damming of the Fraser became much less likely because of the development of the Columbia and Peace Rivers. But other fish vs. power debates emerged on the river in the 1970s and beyond. In the late 1990s, when I initiated this study, Alcan's Kemano completion project produced fraught political debate. Thorough study of these later conflicts will await the release of important archival documents. I hope that my study at least sets down some criteria by which to consider these later disputes and establishes the necessary contexts to understand their dynamics.

Fish versus Power tells the story of one river for which the momentum of modern industrial development to consume available energy supplies did not follow the familiar trajectory. In an era when large dams have come under increasing criticism, the case of the Fraser River can remind us how precarious the environmental defense of large rivers is and how a range of contexts – social, political, economic, and cultural – and scales – local, national, and international – must be explored in order to make sense of the politics of large rivers, no matter the location.

A final note: Several of the institutions and government departments discussed in this book changed their names over time. I have tried to flag clear changes in names where possible. Two institutions, however, deserve special mention: First, The British Columbia Electric Railway Company went through several phases in its corporate development that produced different names. In 1926, the British Columbia Electric Railway Company formed the BC Electric Power and Gas Company as a holding company; it was subsequently purchased in 1929 by the BC Power Corporation. In 1946, BC Electric Power and Gas changed its name to BC Electric Company. No matter the period, public discussion of this corporation often used both names interchangeably. For simplicity, I refer in Chapter 2 to this corporation only as the BC Electric Railway Company (by the acronym BCER) and thereafter as BC Electric. Second, what is now known as the Department of Fisheries and Oceans has had different names over time: From the 1890s to 1930 it was called the Department of Marine and Fisheries; between 1930 and 1969 it became the Department of Fisheries. I refer to this department as appropriate. Fourth, in subsequent chapters, British Columbia will be abbreviated to BC. Finally, all dollar figures given in the text are in Canadian dollars unless otherwise stated.

1

"A Rock of Disappointment"

Stories of the disaster always went back to the fish: pools of them, red, mature sockeye, writhing in the river, penned in and constrained. They could be seen for miles downstream. At Hells Gate in 1913, the salmon migration to the upper basin had been virtually blocked. (I have followed the BC gazetteer's spelling of Hells Gate, despite the widespread use of an apostrophe – as in Hell's Gate.) This place was a narrow passage, difficult for migration in any year. In the summer of 1913, it was worse yet because a series of landslides had changed the river's course, filled pools and eddies, and created impassable falls. The gate had practically turned into a dam. John Pease Babcock, BC's assistant commissioner of fisheries, said it was "a wonderful sight."[1] From 1911 to 1913, railway construction crews had triggered these slides while laying track on the river's east bank. For two years, government construction crews attempted to remove the debris. At Hells Gate, the slope continued to slide.

The scale of the Hells Gate slides and their effects are difficult to imagine. The Fraser was the most productive salmon river in the world; 1913 was one of the spectacular "big years" – when dominant runs, occurring once every four years, return to the river. The slides not only helped to destroy the big-year run of 1913, but also initiated a chain reaction in subsequent salmon runs. Even as the river was modified after 1913, the Hells Gate passage continued to affect the success of salmon runs. In 1917, the first big year after the slides, the catch dropped to one-fifth that of 1913; four

1. Babcock, "The Spawning Beds," p. 22.

years later, it was but a thirtieth of the 1913 level.[2] In combination with fishing pressure, habitat destruction, and changing oceanic conditions, the slides decimated the Fraser fishery.[3] By 1921, John Pease Babcock was persuaded that, "the Fraser is fished out of sockeye. The big run has been destroyed."[4]

The transformation of Hells Gate bore consequences for the social relations of the Fraser fisheries. Regulators imposed new restrictions on native fishing in the canyon and beyond. Declining stocks undercut the supplies of the commercial fishery in the United States and Canada. Fisheries officials and scientists experienced new challenges to their reputations as experts capable of restoring an impoverished nature. By obstructing the salmon, the slides influenced both the economy and society. Rock was not the only thing to slide. So too did the meanings and uses that humans attached to this place. Ideas and perceptions of this event *slid* between different groups and within them.[5]

Hells Gate: Natural and Cultural History

Located in the Fraser Canyon, Hells Gate is 260 km from the river's mouth. At this place the river is constrained by two subvertical walls of granodiorite. For thousands of years they have narrowed the river, raised its velocity, and made upstream salmon passage difficult.[6] The sheer, imposing walls of the Fraser Canyon rise hundreds of meters above the river at Hells Gate. Between winter low and summer high flows, river levels vary by as much as 35 m. The river at this place is a variable element, recording with its movements the shifts of climate and flow.

Salmon began to pass Hells Gate after the last deglaciation, colonizing the basin from 4000–6000 years ago.[7] Long tongues of ice

2. Rounsfell and Kelez, "The Salmon and Salmon Fisheries."
3. For a nuanced discussion of the many aspects of fisheries depletion, see Taylor, *Making Salmon*, Chapter 2, "Historicizing Overfishing," pp. 39–67.
4. British Columbia. *Report of the Commissioner of Fisheries for British Columbia for the year 1913*, p. 6.
5. New, *Land Sliding*.
6. Moore, "Hydrology and Water Supply"; Northcote and Larkin, "The Fraser River."
7. Groot and Margolis, *Pacific Salmon Life Histories*; McPhail, "The Origin and Speciation of *Oncorhynchus* Revisited."

formed wedges and dams in the upper basin, creating large interior reservoirs and lakes. As these tongues melted and disappeared, the upper basin opened to salmon; the lakes became spawning habitat. As with other falls and rough sections on the river, salmon encountered difficulty at Hells Gate; while passing, they hugged its margins and rested in back eddies.

Since the retreat of the last ice sheets, native peoples have lived with the river and the canyon.[8] In BC's southern interior, archaeologists find evidence of nomadic occupation by deer and elk hunters around 7000 years ago and then note a transition to cultural groups that bear markings of coastal influences (the Pebble Tool Tradition) over the next 2000 years. These human migrations into the interior followed on those of salmon. Around 4000 years ago, winter pit houses began to appear on the Fraser, as they did on the upper Columbia, marking increased cultural complexity built on an elaborated salmon economy. Isotopic analyses of human skeletal remains in the interior suggest that groups with access to the fishery obtained from one-half to two-thirds of their dietary protein from salmon.[9] Salmon was by no means the only basis of early human occupation of the canyon and the Interior – hunting and gathering were important as well – but it was an essential staff of life.

Hells Gate was one of the many fishing stations that dotted the canyon in the precontact and postcontact periods. The earliest photographs of Hells Gate show the wooden drying racks of the Nlaka'pamux (Thompson Indians), bearing loads of salmon (see Photograph 1).[10] This station, like other prized fishing places, afforded excellent opportunities to catch fish hugging the river's edge in their attempts to escape rough water. It also offered fish that had lost part of their fat content and could be preserved well for winter storage. Standing on well-appointed rock or wooden platforms that hung by rope from rocks and cliffs, fishers employed long dip nets that cinched their catch.[11] In the late nineteenth and early

8. This summary of prehistory draws on Carlson, "The Later Prehistory of British Columbia."

9. N. C. Lovell et al., "Prehistoric Salmon Consumption in Interior British Columbia."

10. BC, Archives and Records Service (hereafter BCARS), Photograph Collection. Photo A-05620 "Indian Fishing Place, Hell's Gate River" (189–), photographer undetermined; A-03874, Hell's Gate ca. 1867. "Hell's Gate Canyon or the Great Canyon, 23 miles above Yale," photo taken by Frederick Dally (1838–1914).

11. Laforet and York, *Spuzzum*, pp. 60, 69; Teit, *The Thompson Indians*, Vol. 2, pp. 249–50, 293–4; Wyatt, "The Thompson." For a close study of the related fishing practices

Photograph 1. Hells Gate, ca. 1867. "Hell's Gate Canyon or the Great Canyon, 23 miles above Yale."
(Photo taken by Frederick Dally (1838–1914). BC Archives and Record Service #A-03874.)

twentieth centuries, Salishan groups moved up the river to fish in the lower canyon, whereas Nlaka'pamux from the canyon, the Nicola valley, and from as far away as Kamloops and Williams Lake fished around Hells Gate and throughout the canyon and as far north as the Bridge River rapids.[12]

According to the Nlaka'pamux origin tale of salmon, the transformer and trickster figure Coyote created Hells Gate and the

of the Lillooet Indians, see Romanoff, "Fraser Lillooet Salmon Fishing"; see also Kew's excellent overview of the cultural implications of the resource in the same collection, "Salmon Availability."

12. National Archives of Canada (hereafter NAC), RG 23, Vol. 679, File 713-2-2[8]. H. Graham, Indian agent, Lytton, BC, to Duncan Campbell Scott, deputy superintendent general of Indian affairs, November 7, 1925 (Graham here reports of groups coming from more distant locations, like Kamloops, but holding kin ties to fishing sites); Harris, "The Fraser Canyon Encountered," in *The Resettlement*, p. 108; Laforet and York, *Spuzzum*, p. 137.

canyon.[13] Recorded by ethnologist James Teit at the turn of the century, this story finds close parallels in the traditions of various culture groups of the Fraser and Columbia plateau regions.[14] In earlier times, the story goes, peoples who lived at the mouths of the Fraser and Columbia Rivers constructed dams and kept salmon as their prisoners. In those days, the people of the interior did not fish. Coyote decided to change this. He disguised himself as a piece of wood and drifted down the river. When he reached the river's mouth, he broke the dams and guided the salmon to tributaries and lakes in the interior. Later he did the same on the Columbia River. Finishing his work, Coyote declared "henceforth salmon should ascend into the interior each year."[15] With the remains of the broken dams, he forged rocks and made canyons on the Fraser and Columbia. Hells Gate was part of Coyote's work.

This gorge impressed the first Europeans to encounter it, not for the excellent fishing it afforded, but for the perils it created for travel. Hells Gate enters the European historical record with Simon Fraser's account of his 1808 journey to the Pacific. Passing down the canyon, Fraser was disturbed by the roughness of the water and decided to walk this portion of his journey[16]:

> I have been for a long period among the Rocky Mountains, but have never seen any thing equal to this country, for I cannot find words to describe our situation at times. We had to pass where no human being should venture. Yet in those places there is a regular footpath impressed, or rather indented, by frequent traveling upon the very rocks.

In 1859, Commander R. C. Mayne described his travels past Hells Gate, confessing that the territory "makes one's nerves twitch a little at first."[17] Like Fraser, Mayne depended on native paths and rope bridges through the rough sections of the canyon. In the 1820s, Hudson's Bay Company traders estimated the population of the

---

13. "Coyote and the Introduction of Salmon."
14. Laforet and York, *Spuzzum*, p. 37. Laforet and York mention that most of the work for Teit's mythology was conducted in 1898. On other tellings of the Coyote story, see Taylor, *Making Salmon*, p. 29; Kennedy and Bouchard, "Stl'atl'imx (Fraser River Lillooet) Fishing." On James Teit's career and political activities, see Campbell, "'Not as a White Man, Not as a Sojourner'"; Wickwire, "'We Shall Drink from the Stream and So Shall You.'"
15. "Coyote and the Introduction of Salmon," p. 303.
16. Lamb, *Letters and Journals of Simon Fraser*, p. 96.   17. Quoted in Ibid.

canyon at around 7500 people between Yale and just south of
Boston Bar, a 35-km distance. It is likely that this population was
in a period of demographic rebound after the devastation by smallpox epidemics in 1782.[18] Over the next century, Europeans would
follow Fraser and treat the canyon and Hells Gate as a point of
passage, a corridor en route to the Pacific or the Interior.

For the most part, the canyon route passed by, but not through,
Hells Gate. Water transportation, as Fraser had discovered, was
virtually impossible. In 1882, the Hudson's Bay Company sent a
steamer named *The Scuzzy* through Hells Gate, but it required
the force of 150 Chinese immigrant laborers, straining on the
canyon walls, to pull the boat through. The feat would never
be attempted again.[19] During the 1858 gold rush to the upper
Fraser Basin, native and fur-trade routes were transformed into
a passage for miners; later the Royal Engineers turned them into
the Cariboo Road. Across the river, the Canadian Pacific Railway
(CPR) marched past the Gate in the 1880s, connecting the canyon
with the coast and transcontinental markets. Now, not only the
river carried salmon. Railroad cars hauled boxes of canned salmon
to eastern and transatlantic markets. With each progressive connection, the friction of distance was reduced and new agents of
change were introduced to the canyon and its peoples: land surveyors, missionaries, miners, and settlers. The infrastructure of white
Canadian society and waves of American, European, and Asian
immigrants transformed this place and its connections with the
world.[20]

In the early 1910s, the Canadian Northern Railway (CNR), the
second transcontinental, cut into the canyon's walls. The Railway's
financial difficulties at this date, Ted Regehr explains, led to "hur-
ried construction" and a lack of "due caution" by the contracting
firm, the Northern Construction Company.[21] Crews cleared their

18. Harris, "Voices of Smallpox Around the Strait of Georgia" and "The Fraser Canyon
Encountered," both in *The Resettlement*, pp. 3–30 and 105–7, respectively; Wilson Duff
offers a population figure for all Interior Salish (Thompson, Lillooet, and Shuswap) at
13,500 (1835), 5800 (1885), and 5348 (1890). 1890 stands as the lowest point of
Interior Salish population levels: Duff, *The Indian History of British Columbia*, p. 39.
19. Pretious, "Salmon Catastrophe at Hell's Gate," p. 17.
20. On all of these themes, see Harris, "The Fraser Canyon Encountered," in *The Resettlement*, pp. 103–36.
21. Regehr, *The Canadian Northern Railway*, pp. 391–2.

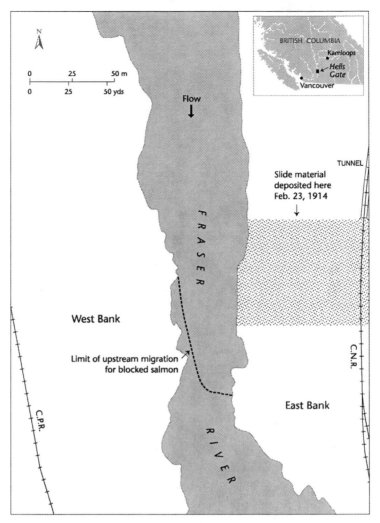

Map 2. Hells Gate and the landslides.
(*Source:* Map in Jackson, *Variations in Flow Patterns*, p. 89.)

path with dynamite and left the rock and debris to fall down the bank and into the river. At Scuzzy Rapids, at China Bar, and especially at Hells Gate, the land began to slide (see Map 2 and 3 and Photograph 2). Here the railroad not only transformed relationships of time and space, but also modified its physical surroundings.

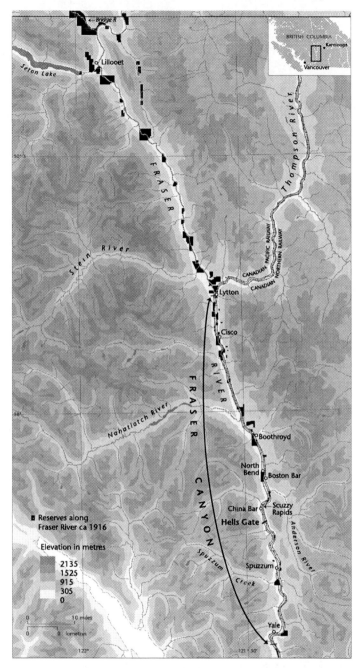

Map 3. The Fraser Canyon, ca. 1916.

26

Photograph 2. "Rockslide, Fraser River at Hells Gate."
(*Source:* "Rockslide, Fraser River at Hell's Gate," March 2, 1914. Photographer undetermined. BC Archives and Record Service #A-04680.)

With cruel irony, Coyote's legacy was filled with rocks and earth. The dams he broke were remade.

"A Milling Mass of Sockeye"

The effects of the slides on salmon migration came into view in the summer of 1913. During the big year of sockeye migrations, the slides acted as an enormous dam. Water found its way through and over the debris. But for the salmon, there was little prospect of passage to spawning grounds.

BC's Assistant Commissioner of Fisheries John Pease Babcock left the best account of the discovery of the slides. Babcock was not a scientist proper, but in the fluid professional world of early twentieth-century fisheries science, he was widely hailed as a fish expert. He had worked earlier in his career with the California Fish and Game Commission and had learned the techniques of fish culture and hatchery development. In BC, beginning in 1902, he opened a small hatchery at Seton Lake to augment earlier federal hatcheries on the Fraser. He also toured many of the principal rivers of the province each summer to assess the state of the spawning grounds. By 1913, he had toured the canyon approximately ten times and was familiar with the place. His understanding of salmon migration rested on a kind of hybrid knowledge that integrated the views of native peoples and other locals he encountered on his tours with contemporary scientific ideas – mainly originating from biologists at Stanford University – about Pacific salmon. These diverse ideas were not always compatible. For many years, for example, Babcock denied popular belief that salmon return to their natal streams to spawn and followed Stanford biologist David Starr Jordan in insisting that this could not be so. Only later did Babcock accept the so-called home stream theory, when Charles Gilbert, another Stanford zoologist, provided strong evidence based on research in BC to defend it.[22] Babcock's optic on the slides and their consequences therefore must be acknowledged as the fragmentary and conditioned lens that it was. It refracted relationships of power and integrated and selected information according to prescribed notions of expertise and relevance.

Early in August 1913, Babcock approached the canyon during his annual tour of the spawning grounds. Advance reports suggested trouble ahead; obstructions were apparently blocking fish.[23] Babcock rushed to the scene. At Hells Gate he stood on the cliffs above the river and observed numerous fish, milling in eddies below the passage. He could also see some salmon swimming through, but water levels were high and the river a muddy brown. Activity beneath the water surface was difficult to discern. Moving about the canyon, Babcock spoke to natives who claimed that they had caught few fish above Hells Gate since mid-July. At Seton Lake, the site of Babcock's hatchery, only a thousand fish had returned. This was well below expectations. He wondered hopefully whether fish were passing below the surface, beyond his obscured view. Four days later, accompanied by Stanford zoologist Charles H. Gilbert, Babcock walked the banks of the river through the canyon. Above Hells Gate, every eddy before the Scuzzy Rapids was filled with a "milling mass of sockeye" (see Photograph 3). Not many fish seemed to be passing Scuzzy Rapids. Returning south, Babcock and Gilbert found a now-familiar sight at Hells Gate: "Vast numbers were seen approaching the Gate on both sides of the channel.

22. On the nature of Gilbert's contribution to fisheries science, see Smith, *Scaling Fisheries*, pp. 28–33.
23. My account of Babcock's journey to the canyon is based on Babcock, "The Spawning Beds," pp. 20–38.

Photograph 3. Sockeye Salmon in Spuzzum Creek, August 1913.
(*Source:* "Sockeye Salmon in Spuzzum Creek, August 1913." Photographer undetermined.
BC Archives and Record Service # G-02235.)

They filled every inch of space where they could make headway against the stream, and even in the most rapid parts of the channel fish were seen struggling to advance."[24]

The next day Babcock began to talk with local residents to see what light they could shed on the matter. William Urquhart, a track watchman with the CPR, living in Spuzzum, said that there were more fish in the canyon than he had seen in twenty years. James Paul, chief of the Spuzzum band, said that salmon always massed in great numbers in August and September of the big-run years. But he too could not recall a year with so many fish, except a time "many years ago." Henry James, a native of the canyon remarked that, "all the old Indians remember only one other year, many many years ago, when the salmon had been so thick as this year." Edward Farr, a CPR masonry inspector, said that it was true that salmon massed in the canyon in a big year, but that the CNR's construction

24. Ibid., p. 22.

had thrown rocks above and below Hells Gate, filling resting bays for salmon and changing the current. Other CPR employees, D. Creighton and Thomas Flann, recognized that there was a holdup, but "believed that all would pass through in time."[25]

But would they? Babcock went north to investigate. Initial reports from the Chilcotin and Quesnel Rivers suggested good returns as in previous big years. But the arrival of fish in mid-August and late August at these points was followed by mysterious weeks of no fish. James Moore, a BC Department of Fisheries watchman at the Quesnel dam, reported spotty results. Weeks of healthy returns had been followed by weeks of low returns. Native peoples fishing on the Chilcotin similarly found fish for a time, but then reported that the returns had fallen off in early September. The patterns, if there were patterns, did not make sense. In a big year, returned migrants to the upper basin should have been uniformly high in number. Babcock surmised that there must be serious problems in the canyon. He abandoned any further investigations in the upper basin and returned to Hells Gate on September 18.

What had changed? Water had dropped by about 15 ft, but the picture of salmon milling in eddies remained. With the water lower, the full extent of the rock debris was becoming more visible and, it appeared, more difficult for fish to surmount. Babcock abandoned his earlier hope of subsurface passage and acknowledged that fallen rock had obstructed salmon. He began to formulate emergency restoration plans. Rock would have to be cleared quickly at Scuzzy Rapids and at Hells Gate. Babcock did not yet realize that problems existed at China Bar and White's Creek as well. As the water dropped, these problems would rear up visibly against the flow. But for now he decided explosives were the thing: Clear the debris with some well-placed dynamite at the two worst points and hope for an easier passage. Having worked independently of any federal officials to this point, Babcock now passed word of the problems to officials in Victoria, and the Department of Marine and Fisheries in Ottawa was asked for support. By September 28, a more effective passage had been opened, and the matter seemed in hand. Work ceased, and officials of the federal and provincial

25. Ibid., pp. 23–4.

governments departed. Babcock returned on October 10, only to find that the water level had dropped again and the former artificial passages lay dry. Blaming the difficulty on a native employee he had left in charge of the site, Babcock took pains to disassociate himself from the difficulty in a subsequent report. The situation was as bad as it had been at any point in the season. Babcock secured men from the CPR, wired for help, and then set clearance activities in motion again. Later in the month one federal official would write to his superiors of their labors at "that hated place Hells Gate."[26]

To this point, the clearing of the obstructions had been haphazard and the effects of the rockslides unclear. Now, with a sense of urgency and official support, clearing of the rock debris began in earnest. For two months, G. P. Napier of the Provincial Public Works Department oversaw operations, but was replaced in December with J. H. McHugh, an engineer of the federal Department of Marine and Fisheries.[27] Work began at Hells Gate and Scuzzy Rapids and entailed the planting of sticks of dynamite (usually forty at a time) in wedges between the boulders. As the rocks blew apart, masses of rotting flesh and fish bones flew into the air. Some provincial fisheries staff collected exhausted fish below the blockade, slit them open, and took their spawn to provincial hatcheries.[28] Crews were sent into the woods to hew timber to produce towers on either side of the river to support a crossing of cables that would carry a shovel to dig out the mess from above. Through the winter, progress seemed steady. The water had dropped. Conditions were easier, if colder. Sometimes up to ninety men dug, cleared, and passed the rock above the river and across the channel.

And then all the work was destroyed. On February 10, 1914, local CPR workers felt a steady shower of small rocks during the day; that night, around 10 P.M., a huge section of the cliff, and

26. NAC, Pacific Region Office, RG 23, Vol. 2307, File 1–11, H. Walter Doak to A. P. Halliday, assistant inspector of fisheries, October 22, 1913.
27. Napier wrote a short account of the clearing conditions: Napier, "Report on the Obstructed Conditions of the Fraser River," pp. 39–42. McHugh's detailed discussion of the clearing activities is contained in McHugh, "Report on the Work of Removal of Obstructions," pp. 263–75.
28. UBC, International Pacific Salmon Fisheries Commission Papers, Canners' Scrapbooks (hereafter CSB), "Clear Blockade in River Canyon," *Province*, January 23, 1914.

a 50-ft section of the CNR railway bed fell into the passage at Hells Gate. Railway tracks lay by the water's edge, "twisted into all shapes."[29] The river was again remade.

When McHugh assessed the damage the next day, he tried to gain a sense of the width of Hells Gate under the press of the new material. He tossed a rock attached to a line across the passage and measured it at around 75 ft. Of those 75 ft, 15 ft were consumed by barely covered debris underwater. Before the slides, it is estimated that the passage measured 110 ft in width. The velocity of the current passing the Gate was intense; McHugh and his assistants lost their gauges when trying to measure it. The slide had "practically formed a dam," wrote federal Chief Inspector F. H. Cunningham. And the problems, he continued, were new: "[T]hese rocks will change the currents, and will, of course, obliterate the eddies which existed, and which assisted these fish in their ascent of the river. In fact, one might say that the whole character of the river at this point is changed, and this slide has created an entirely new problem for consideration."[30]

Standing on top of the ruins of the railroad, examining the scene, McHugh understood well enough the difficulties ahead, particularly in view of the fact that the salmon migrations would be on them again by July. After the CNR's reluctant response to the federal department's demands for assistance, the Pacific Coast Dredging Company was engaged to assist McHugh. What had worked before was tried again. Explosives and the shovels loosened material underwater, and the force of the current carried it downstream. Similar methods were used at Scuzzy Rapids. By March, as interior snowpacks began to melt, the water level steadily crept up the Gate's wall, 1 ft a day, in McHugh's estimation. Ninety men cleared and fought against the rising current at the busiest times of the season, trying to ensure as much clearance as possible before the arrival of salmon (see Photograph 4).

And then they came. On July 3, salmon were observed trying to pass Hells Gate. "The most interesting part of the work," McHugh recalled, "was at hand."[31] For over a week, there was no sign of

29. NAC, Pacific Region Office, RG 23, 2311, File 5–2, Cunningham to W. A. Found, superintendent of fisheries, February 18, 1914.
30. Ibid.   31. McHugh, "Report on the Work of Removal of Obstructions," p. 270.

Photograph 4. Clearing the Gate, 1914.
(*Source:* "Men working on left bank to remove slide Hells Gate April 13, 1914." Photographer undetermined. Pacific Salmon Commission Collection.)

fish passing Hells Gate. On July 15, three were caught above the Gate. But the massing of bodies below the rough water was becoming evermore noticeable, and, when contrasted with the meager numbers above the Gate, called for attention. Three natives were employed for a number of days, dressed in oil skins and sou'westers, and instructed to catch fish in the rough water and eddies by using dip nets.[32] Cinching the fish, they would turn and place them in wooden flumes to swim up and around the obstruction. They may have assisted up to 20,000 fish in this manner. Other flumes, jutting into the rough water, were set up by men hanging from ropes, "drenched in icy spray" in order to steer the fish through a 300-ft diversion.[33] Some fish entered; most did not. Below the crews, the fish swam in a circle: some swam up along the bank on the CPR (west) side of the river and passed through; many others followed the west bank and then tried to cross the river and pass on the east side, facing an intense current. Most of these fish were swept back down the river, hit an eddy, and were brushed yet again into the throngs of salmon attempting passage on the west side. "I can testify," wrote one CPR employee who viewed this struggle, "that the fish were doing their best, throwing themselves out of the water in their eagerness and wounding themselves against the rocks or the side to get a purchase against the current."[34] By mid-August, it appeared that most of the fish trying to do so had passed. But then, many had probably floated down river, unspawned, as they had the year before. Babcock judged at the end of the season that few had reached the spawning grounds. "I ran up to Adams Lake," he wrote to Charles Gilbert, the Stanford zoologist, "and the Indians there tell me they had taken but six fish this year. I could find none in the Adams River. At Seton Lake we have about 150 fish. You will admit that the prospect is a poor one."[35] By December, after the last fish had arrived, much of the clearing work was

---

32. NAC, Pacific Region Office, RG 23, Vol. 2307, File 1–18, McHugh to Cunningham, August 23, 1914.
33. CSB, "Work on Hells Gate Described," *Province*, April 16, 1914.
34. NAC, RG 23, Vol. 678, File 713-2-2[2], William P. Anderson to deputy minister of marine and fisheries, September 26, 1914.
35. BCARS, GR 435, BC Department of Fisheries, Box 56, File 510, Babcock to Gilbert, August 29, 1914 (copy). Babcock continued his dismal assessment in a series of letters to Gilbert in this file.

complete, and, with the water lower, the site could be inspected at the different points of difficulty.

The results were reassuring. The force of the river seemed to have carried away many of the large boulders at China Bar and Scuzzy Rapids, previously weakened by explosives. The fall at Hells Gate, measuring 5 ft before the slides, was now at about 9 ft. It had stood at 15 ft at the worst moments before the clean up, and 6 ft of progress seemed like a lot to McHugh, although it would mean an extra 4 ft for passing salmon. One man had died, and four were injured in the clearance work. This seemed like a good result to the engineer in charge. The cost of all the work totaled $110,212.70, later to be paid by the CNR by means of a reduction in a government grant to the company, after the railway balked at the expense and an outright payment. At last it could be said that the matter was finished.

Encountering an environment without fixed meanings, Babcock and McHugh explained their experiences in narratives of discovery. They ordered their subjects through measurement: Babcock tried to account for salmon numbers, McHugh assessed distances and gauged velocities. They gathered local knowledge and interwove it with established principles of biology and engineering: Babcock interpreted interviews with locals to produce meaningful signs of change and discord; McHugh adopted native fishing methods to help fish pass around the slides. Like discoverers, they took possession: Both Babcock and McHugh ordered the landscape and assumed to know how to correct it; they held the authority to control space and resources and used it. Although the event and immediate aftermath of the Hells Gate slides occurred under a cloud of confusion, the authors of the accounts, which provide the backbone of the preceding narrative, imposed sets of meanings that can be considered only tentative and partial. From a certain perspective, their emphasis on the unknown served to excuse their inability to identify and correct the problems sooner: The accounts contain an embedded rhetoric. Unfortunately, it is difficult to balance their descriptions and judgments against others; they are the only authors of substantial contemporary accounts of the slides and clearing activities. To begin to understand the flow of meanings through Hells Gate in 1913–14 and then the continuing stream of questions and arguments that followed in their path, it is necessary to establish contexts for understanding the significance of the slides for native peoples, the commercial fishery, regulators, and scientists.

The Native Fishery

In the midst of clearing the Hells Gate slides, J. H. McHugh was irritated by the arrival of native peoples, preparing to conduct their traditional fishery. In July 1914, while McHugh and his teams labored to excavate debris and blow up portions of the passage, pony trains ambled down the hillside, some from as far away as the Nicola Valley. The fishery of the Nlaka'pamux was about to commence. McHugh understood that the fishery at Hells Gate dated to "time immemorial," but as the natives unpacked their gear and readied themselves for the arrival of the salmon, he became convinced that a "wholesale slaughter" would ensue. He reasoned that all the restoration work would be for naught if the fishery were allowed to proceed. In the face of "strong and organized objection," McHugh ordered restrictions placed on the natives and their fishery and assigned special guardians to police the area and enforce the informal order. Why were the natives so agitated? McHugh stated that it was possibly "the first time this ancestral privilege had been in any degree interfered with." He felt sure, nevertheless, that in the moments when fishing was allowed that season, the natives "doubtless received all the fish they required." It was a self-serving judgment, and natives of the canyon disagreed.[36]

After the slides and the imposition of a new regulatory presence, natives organized their defense. By the end of July, after having missed almost a week of prime fishing, a number of chiefs in the canyon made common cause to publicize their predicament and force the Department of Indian Affairs to correct the matter. While the local Indian Agent attempted to broker a solution with Assistant Chief Inspector A. P. Halladay of the Department of Marine and Fisheries on July 25, natives developed another strategy.[37]

36. McHugh, "Report on the Work of Removal of Obstructions," p. 271.
37. CSB, "Indians Resent Fish Embargo," *Columbian*, July 25, 1914. In private correspondence, Halliday wrote that H. Graham, the local Indian agent, agreed that Indian

Bypassing their state guardian, Chiefs James of Yale, Michael of Maria Island, Paul of Spuzzum, and Jimmy of Ohamil wrote to the department:

We the representative of the tribes of Indians between Hope and Lytton wish to notify you that the fishery department of New Westminster have stopped us from catching salmon in the Fraser River, for our own use. This we refuse to do. There is no sense of justice in this order, as all the fish we Indians would catch in the year would not equal the number caught in one day by the white men at the mouth of the river. We Indians wish to tell you that the way to save the fish is to stop the white men from setting traps and nets, so blocking the mouth of the river that the fish cannot get up. The white man [sic] are to blame for the scarcity of fish, and yet they would take away from us Indians the only means of making a living after taking everything back from us. This we positively refuse to submit to, and look to you for justice. We have now been stopped six days and we expect damages for this delay.

"[W]e are the original owners of the land," they continued, "and we know more on the fish than any individual or government."[38] After three days without reply, the chiefs telegrammed again on July 29, demanding action.[39] The following day the department responded that it had notified the Department of Marine and Fisheries about the matter and that this department would send an official to investigate and report. No action would be taken until the said report was received. The chiefs, obviously irritated at this treatment, promptly leaked the correspondence to the press. It was reprinted in full in the *Vancouver Sun*. It was not until the end of the month, after federal Fisheries Officer F. H. Cunningham had surveyed the scene, that it was decided to change the restriction to a four-day-per-week fishery.[40] Cunningham admitted on his return to Vancouver that natives were indeed "aggrieved" by this decision but that they were sticking to it.[41] After investigating their legal options and engaging the firm of Harris, Bull, Harrington, and Mason

fishing should be restricted: NAC, Pacific Region Office, Vol. 2311, File 5–7, Halliday to Cunningham, July 28, 1914.

38. CSB, "Indians Determined to Get Fish Supply," *Vancouver Sun*, August 4, 1914.
39. Ibid.
40. NAC, Pacific Region Office, RG 23, Vol. 2307, File 1–18, Cunningham to Found, August 13, 1914.
41. CSB, "Much Debris Still to Move," *Columbian*, August 24, 1914.

to press their claims with the Department of Marine and Fisheries in Ottawa, natives were faced with little choice but to accept the new restrictions or face arrest.[42]

The crisis at Hells Gate triggered changes in native fishing rights in the canyon. However, it paralleled a more general shift in the regulation of the native fishery in BC that spelled increasing restrictions and a decreasing native catch. In the pioneer period of the commercial fishery, natives worked in and fished for canneries and maintained their own food fishery without significant interference. But with the sharp expansion of canneries around the turn of the century, native involvement in the commercial industry declined (they were frequently displaced by immigrant Japanese fishers), while commercial pressure on the government to restrict the native food fishery grew.[43] In an environment of decreasing supply and an expanding commercial fishery, native uses of salmon were equated with waste. Natives became portrayed as depraved animal killers, destroying a resource that could be profitably put to use. Some commentators suggested that a ration system for natives would be better than preserving their access to fish.[44]

Across the province, a series of confrontations after 1900 marked the introduction of the new regulatory regime. Using previously ignored sections of the Fisheries Act outlawing obstructions on rivers and streams, fisheries officials cracked down on native fishing traps and weirs across the province, at the behest of local canners. At Babine Lake, on the upper Skeena, there was a major standoff in 1906 over the right of natives to employ fishing weirs. It led to a series of arrests and prosecutions, sensational metropolitan newspaper coverage (the event was dubbed the Babine "uprising"), and finally a trip by chiefs and a local Oblate missionary to Ottawa to meet with Prime Minister Laurier to resolve the dispute.[45] Other similar incidents occurred on the Cowichan River in 1897, 1908, and 1912, and on Clayquot Sound in 1906.[46] In another dispute,

42. NAC, Pacific Region, RG 23, Vol. 2307, File 1–18, Harris, Bull, Harrington, and Mason to Halliday, August 18, 1914.
43. Newell, *Tangled Webs*, pp. 46–50; Ray, *I Have Lived Here*, pp. 296–8, 302.
44. CSB, *Vancouver News-Advertiser*, September 23, 1906, no title.
45. Harris, *Fish, Law, and Colonialism*, pp. 79–126.
46. CSB, "Indians Barricade Cowichan River," *Province*, May 30, 1908; Newell, *Tangled Webs*, p. 90.

over the use of fishing traps near Salmon Arm, a group of fifty natives sprung two arrested chiefs from court.[47] The shift in the regulatory regime produced resistance.

Conflicts over fishing rights were part of a more general struggle over the land question in BC. After 1900 increasing settlement and the expanding white resource economy placed intense pressures on native societies. As Dianne Newell argues, fishing rights held a special position within the debate over resource rights because they underpinned such a significant aspect of native economy.[48] Spurred on by conflicts like those already discussed, native peoples created local groups to defend their interests, such as the Nisga'a Land Committee and the Interior Tribes, and also forged a pan-regional organization, the Indian Rights Association. These groups introduced white society to a new generation of mission-educated leadership with the political skills to engage the settler society and a practical concern to maintain aspects of traditional livelihood.[49] At the time of the Hells Gate slides, natives in the canyon were involved in the broader political debate and were part of a lobby to press for a resolution of resource concerns.

One forum for this struggle, in part created in response to native pressure, was the joint provincial–federal McKenna–McBride Royal Commission of 1913.[50] Intended to vent native frustrations and reassess previous reservation allotments, the commission toured the province in 1913–14, interviewing natives and Indian agents and surveying native communities. Narrowly conceived in the view of many native protesters, the commission was boycotted by some groups for excluding more fundamental questions of land title. In the fall of 1914, a few months after the dispute between native groups and the Department of Marine and Fisheries, the commission toured the Fraser Canyon. Not surprisingly, the commissioners heard more than they might have expected about fish. Representatives of the Boothroyd, Cisco, Spuzzum, North Bend, and Yale bands all raised the problem. The question put to the

47. CSB, "Indians are Troublesome," *Vancouver News Advertiser*, August 18, 1908.
48. Newell, *Tangled Webs*, p. 3.
49. Galois, "Indian Rights Association"; Tennant, *Aboriginal People*, pp. 84–95.
50. On the role of native pressure in forcing some form of response, see Galois, "Indian Rights Association," pp. 16–19.

commissioners by Chief Paul Heena of the Spuzzum band aptly summarizes the concern:

Whose fault was it that I hadn't sufficient food to eat this year? Who was the cause of our poverty? It was not my fault that today we are poor. I was stopped from providing myself with food. No one should be stopped from providing themselves with food. When they came to stop me they told me that if I did not obey I would be put in gaol.

One of the commissioners pointed out in reply that the slides originated from a variety of causes and that fish needed protection. Heena countered, "The reason of this slide was caused by the white man." Commissioner McKenna observed that the slide "was not an act of man – it might have happened if the white man had never come to this country." Another speaker, Patrick of Boston Bar, insisted that "God Almighty put me here..." and suggested that it was not God's will to impose restrictions on native uses of the lands and rivers. "...I don't want to be stopped from fishing salmon in the River. God made those for our use, and it is from salmon that I make my living. Therefore I wish everything to be set free."[51] In a world of increasingly circumscribed economic roles, limited lands, and new forms of regulation, the slides at Hells Gate and the fishing restrictions devised to help correct them appeared to native residents of the canyon as injustices meted out on innocent bystanders.

Nor did natives find much to admire in the specific measures chosen to restore the fishery. As a previous quotation makes clear, there was a strong sense among native peoples that "we know more on the fish than any individual or government." Such sentiments were not new. Seven years previously, St'at'imc (Lillooet Indian) chiefs had criticized the Seton Lake hatchery as a destroyer of salmon populations. After marshaling their case in the Vancouver *World*, they asked rhetorically, "And now is there anybody who dares to say that we the Indians are the cause of the disappearing of the salmon?"[52] Disregarding the authority of nonnative

51. Commission testimony may be found in BCARS, GR 123, Canada Department of Indian Affairs, BC Records, Vol. 11025, File A-H-7; quotations are from pp. 127 and 275. Cole Harris considers this testimony within the broader context of encounter in "The Fraser Canyon Encountered," in *The Resettlement*, p. 134.
52. CSB, "Chiefs Write Letter to the World," *The World*, October 12, 1906.

salmon experts, they identified the instruments of scientific salmon production as harmful.[53]

At Hells Gate, following the slides and their supposed correction, native inhabitants of the canyon again cast a jaundiced eye on the practical abilities and knowledge of fisheries officials. In the fall of 1916, as McHugh oversaw some continuing work on Hells Gate, he was confronted by what he called a "deputation" of Indians from the Lytton and Nicola Districts, eager to discuss the department's previous work and suggest improvements.[54] In particular, they believed that the removal of a key rock in the Gate would help passage significantly. McHugh dismissed the plan, believing it would intensify the flow and remove certain resting pools, but the natives were adamant: "They are inclined to criticize the improvement work already done by this Department at Hells Gate, and suggest that the CPR are standing in the way of improvement work at Hells Gate because of the likelihood of destroying the scenic beauty of this place should they be allowed to work out their own ideas. This suggestion was, of course, dismissed as being ridiculous."[55] The native claim is revealing of a strong suspicion that matters of environmental regulation were not set simply by fisheries officials, but related to a complex industrial and aesthetic politics. The practical aspects of the natives' proposal were taken with a degree of seriousness, or at least according to procedure. F. H. Cunningham forwarded a photograph to department officials in Ottawa showing a group of natives by the rock they wished to remove, accompanied by a letter from Chief Benedict of the Boothroyd band.[56] On reflection, the department rejected the suggestion.[57] The local Indian Agent, admitting that he had been under pressure to gain permission for the rock removal for some

53. CSB, "Seaton [sic] Lake Hatchery a Miserable Failure," *The World*, n.d., but probably October 1906; Drake-Terry, *The Same as Yesterday*, pp. 215, 224; Meggs, *Salmon*, pp. 82–3. Interestingly, later fisheries scientists would agree that the hatchery work at Seton Lake had failed and may have done long-term damage to local stocks: Thompson, *Effect of the Obstruction*, pp. 59–61.

54. NAC, RG 23, Vol. 678, File 713-2-2[6], J. H. McHugh to F. H. Cunningham, October 16, 1916.

55. Ibid.

56. NAC, RG 23, Vol. 678, File 713-2-2[6], F. H. Cunningham to W. A. Found, October 20, 1916. I have been unable to locate this photograph.

57. Ibid., Found to Cunningham, October 27, 1916.

time, stated that the band would be deeply disappointed; he noted that it was "doubly difficult for the Indians to get their winter supply as of old."[58] Whether or not the specific measures proposed by members of the Boothroyd band would have improved the situation, they were keenly aware, by their own fishing experience, that all was not right with the Gate.

In the years and then decades after the Hells Gate slides, fishing decreased as a contributor to the native economy on the Fraser. In the canyon, fishing regulations and decreased runs reduced catches to the point that one band allegedly offered to sell its fishing rights to the Department of Marine and Fisheries.[59] Such an event was exceptional, but it points to the disastrous effects of the slides. At certain points during World War I, native peoples in the canyon had to rely on rations to survive. Altered fishing regulations in the 1920s made native access increasingly difficult throughout the Fraser Basin.[60] "The salmon question," protested Chief James Paul of Spuzzum in 1922, "is the most important of all things for us. We must have free access to the salmon for our food."[61] At Spuzzum, some natives like Willie Bobb responded to the new state of affairs by finding waged work on the railroad, gardening, fishing, and working at a mining claim; still others retreated into the mountains, hunting "in the old ways as long as they were able."[62] North of the canyon on the Nechako plateau, Carrier groups reoriented their livelihood, hunting more intensively for moose than previously, turning to the Skeena system for fish, and exploiting a rise in fur prices to respond to changing environmental conditions.[63]

58. Ibid., S. Stewart, acting superintendent general: to G. J. Desbarats, deputy minister of marine and naval Service, October 27, 1916, containing enclosure: H. Graham [Indian agent] to Department of Indians Affairs, October 21, 1916.
59. Department of Marine and Fisheries, *Annual Report, 1918* (1919), p. 12, cited in Newell, *Tangled Webs*, p. 117.
60. The Department of Indian Affairs objected to the increased restrictions: NAC, RG 23, Box 679, File 713-2-2[8], H, Graham, Indian agent, Lytton, BC, to Duncan Campbell Scott, deputy superintendent general of Indian affairs, November 7, 1925. Scott forwarded this correspondence to the Department of Fisheries, but the department would not change the regulations.
61. Quoted in Laforet and York, *Spuzzum*, p. 190. This protest recorded the views of Paul and "seven others" as reported by James Teit in a 1922 report for the Department of Indian Affairs on the economic affairs of natives of the Fraser.
62. Laforet and York, *Spuzzum*, p. 106.
63. Hudson, "Internal Colonialism." This is a chapter from Hudson's PhD thesis on the Carrier Indians of northern BC.

The slides produced long-term effects for the native economy and society in the basin.

The Commercial Fishery

Operators in the commercial fishery had a more oblique connection to the Hells Gate slides than native peoples in the canyon and were less affected by its immediate consequences. At the commanding heights of the industry, there was little knowledge of the incident until the story was carried in the metropolitan press in the fall of 1913, and even then there was precious little interest.[64] Yet the record of industry attitudes is revealing. In the fall of 1914, a gathering of the BC Canners' Association – made up of all of BC's important fish-processing firms – with two fisheries officials, federal Chief Inspector F. H. Cunningham, and provincial Deputy Commissioner McIntyre, discussed a number of important issues for the fishing industry. The meeting ended with a series of concluding motions. One of them referred to Hells Gate[65]:

That this meeting heartily endorses the efforts made by the Government for removing slides and other obstructions which prevented salmon from reaching the spawning grounds, and also in preventing Indians and others from taking out fish which were temporarily barred from ascending until such obstructions were removed.

With a bluntness appropriate to the occasion, the cannery representatives established their approval for state actions in protecting their interests against environmental obstructions and native claims. Although overfishing might logically have been implicated in the poor returns to spawning grounds in 1913, it was easier and more desirable to focus attention elsewhere.

Well before the fish struggled at Hells Gate in 1913, they had run a gauntlet of nets. Returning from the ocean by the west coast of Vancouver Island, the majority of fish approached the mainland via Juan de Fuca Strait. Crossing invisible human boundaries, they entered American waters and were caught in fish traps and purse seine

---

64. For one exception to this generalization see Bell-Irving, "Conditions in the Fraser Canyon."
65. UBC, *BC Salmon Canners' Association Minute Book* (March 13, 1914–October 9, 1920), entry for November 25, 1914, p. 21.

nets in high numbers. When Canadian fisheries officials tabulated the pack at the end of the season, it would be shown that Americans, fishing for Puget Sound canners, had reaped about 60% of the total catch.[66] Once back into Canadian waters, fish proceeded to the Fraser estuary, where again gill nets appeared, hanging from the hulls of boats and staffed by native, white, and Japanese fishers. They would be canned in one of thirty-five canneries in the Vancouver region and handled in these places by a predominantly native and Japanese female labor force. The year 1913 was exceptional, one of the famous big years of the sockeye runs, and the canneries overflowed; well over 2 million cases were produced. Reports of the season told of fishers throwing fish overboard because the canneries were ill-prepared for the cornucopia.[67]

The force of this fishery – international, competitive, and largely open – was barely restrained by the state. Despite some notable attempts, no international agreement bound the United States's and Canada's catch levels; with more efficient (and destructive) technology like fish traps and fewer regulations than in Canada, American fishers reaped large returns with minimal concern for conservation. Canadian fishers were only slightly more constrained. Ever since the reception of Canadian fishery legislation after Confederation, the emphasis of federal policy had been to allow the industry to grow, while supporting measures to increase supply.[68] Experiments with limited cannery licenses, weekly closed periods, and gear restrictions helped to impose certain limits, but were frequently ignored by the industry and poorly enforced.[69] The state was more effective when collecting statistics or, as occurred increasingly in the 1920s, excluding Japanese fishers on racist grounds with support of white and native fishers and canners.[70] State-sponsored conservation, such as it existed, operated in the limited field of habitat restoration and hatchery production. Supported by canners in their quest to improve on nature, the federal and provincial

66. See actual figures of the catch in the *Report of the Commissioner of Fisheries for British Columbia for the year 1913*, p. 12.
67. Meggs, *Salmon*, p. 95.
68. Anthony Scott and Philip A. Neher, "The Evolution of Fisheries Management Policy," in Scott and Neher, *The Public Regulation*, p. 11.
69. Newell, *Tangled Webs*, pp. 70–1; Scott and Neher, *The Public Regulation*, p. 11.
70. Newell, *Tangled Webs*, p. 85; Ray, *I Have Lived Here*, p. 305.

departments sponsored hatchery programs and became evermore vigilant in their quest to remove native fishing traps, decrepit mining dams, landslides, and logging slash in smaller streams.[71] Rather than contend with the boiler politics of fishing regulation, fisheries officials used the field of habitat restoration as a regulatory release valve.

Although the Canadian state imposed a light regulatory hand, the industry created its own forms of control on the rapid growth of the late nineteenth and early twentieth centuries. Since the first canneries had been planted on Lulu Island in 1871, the industry had grown to include sixteen canneries in 1890 and forty-two by 1900.[72] In the pioneer fishery, until the late 1880s, these canneries were locally controlled and small in scale. As the fishery developed and new markets for canned salmon of different varieties opened in Britain, the Commonwealth, and parts of Europe, outside capital from the United States and Britain provided the basis for expansion.[73] In an attempt to scale back competition and benefit from economies of scale, BC Packers was formed in 1902, merging twenty-two previous firms and absorbing control of over 50% of the Fraser catch.[74] In terms of regulating the size of the catch, attempts were made at different times to agree on closed periods to allow for fish escapement to the spawning grounds.[75] In a few extreme periods when catches declined sharply, some canners proposed the total cessation of fishing for a four-year period to allow runs to rebuild; with all of the different interests involved, however, and uncertainty as to the practical results of such an undertaking, a closure of this kind never occurred. Less drastically, certain firms sponsored their own hatcheries as a means to supplement their capital stock.[76] Members of the canning interest were thus conscious of the potential volatility of the industry imposed by

---

71. Lyons, "Appendix 17 Salmon Hatcheries Operated in British Columbia by the Dominion Government, 1884–1935," in Lyons, *Salmon*, p. 668; Newell, *Tangled Webs*, p. 52.
72. Lyons, *Salmon*, p. 706. Note that the number of canneries operating fluctuated from year to year, depending on expectations of a big run or a low year. After 1902 and the merger into BC Packers, certain canneries were closed. The number on the Fraser fell to twenty-one in 1910, fifteen in 1911, fourteen in 1912, but then expanded again to thirty-five in 1913 in expectation of a big run.
73. Reid, "Company Mergers," p. 282–302.   74. Ibid.
75. Newell, *Tangled Webs*, p. 72.   76. Ibid., p. 52.

Table 1. *Fraser River Sockeye Catches on the Big-Year Cycle, 1901–33*[77]

Year	Total Catch
1901	25,760,031
1905	20,681,236
1909	20,936,474
1913	31,343,039
1917	6,883,401
1921	1,686,241
1925	1,828,716
1929	2,059,178
1933	2,450,436

Source: Rounsfell and Kelez, *The Salmon and Salmon Fisheries*, pp. 761–2.

changing conditions and unchecked growth in catch capacity. But in a resource economy of lightly controlled access, such concerns, even when backed by powerful cannery interests, could not begin to establish boundaries of conduct.

In a peculiar sense, the Hells Gate episode of 1913–14 bore out the concerns of industry and state officials about the health of inland waters while diverting attention yet again from questions about the fishery. Its results were slow to arrive, as the sockeye spawning cycle operates on a four-year basis. Whereas native fishers felt the slides' effects immediately because of increased regulations and declining returns above the canyon, the commercial fishery experienced a four-year delay. But in 1917, at last, the full effects of the slides reached the cannery interests downstream and with crushing force. Whereas the 1913 catch stood at 2,401,488 cases (Fraser canners took 736,661; Puget Sound canners, 1,664,827), in 1917 it was a paltry 559,702 (148,164 for Fraser canners, 411,538

77. These statistics are taken from the calculation of the combined Canadian and American catch in Rounsfell and Kelez, *The Salmon and Salmon Fisheries*, pp. 761–2. Populations may be estimated somewhat differently from a model of the escapement of salmon to the Fraser Basin. Research by P. Gilhousen suggests that 6,610,000 salmon escaped past the commercial fishery in 1913, 595,000,000 in 1917, and 220,000,000 in 1921. The declining trend after 1913 is less immediate in this model; it contrasts much of the historical evidence of fisheries' conditions and past estimates and must be approached cautiously. See Gilhousen, "Fraser River Stock Escapements."

for Puget Sound canners) (see Table 1).[78] Although Babcock assured the public that the slides had been cleared and the Gate returned to its original condition, the big-year cycle of sockeye appeared to have been destroyed, or at the very least diminished substantially. The *Industrial Progress and Commercial Record*, an optimistic booster of BC business as its title suggests, announced "The Decline of the Sockeye" in August 1917. The Hells Gate disaster was believed to be the major cause.[79] Henry Bell-Irving, a pioneer canner and a prescient observer of the industry, stated before a Royal Commission investigating the fishery in 1916 that the Fraser fishery was "practically a thing of the past." He hoped the same fate would not befall the northern rivers.[80]

The Hells Gate disaster thus helped to reinforce a spatial and species shift in the organization of the BC commercial fishery. Before the slides, growth in the industry was already occurring on new fishing grounds in the North, on the Skeena and the Nass and at Rivers Inlet. After 1913, expansion to these new fields only increased as Fraser stocks declined.[81] The remaining fishery on the Fraser responded to new conditions by expanding the breadth of its reach: Reacting to new markets for pink salmon and other fish species, canneries diversified their product lines. Following a pattern of industrial fisheries worldwide, the Fraser canners responded to species collapse by fishing less lucrative species more intensively and traveling down the food chain in search of new products.

The 1913 big-year catch was the last of its kind. From the vantage point of the coastal metropolis in 1917, the Hells Gate slides appeared as a mysterious interior force, destroying an industry built on the hope of secure supplies. The cannery interest failed to take the opportunity to examine seriously its own role in the downturn, but instead reinforced pressure on government to improve conditions and allow industry a free hand in reorganizing its fishing effort. Responding to market factors and the environmental

---

78. British Columbia, *Report of the Commissioner of Fisheries for British Columbia for the year 1917*, p. 19. This report revised earlier figures for 1913 that were slightly higher.
79. "The Decline of the Sockeye." 80. Ibid.
81. Newell, "Dispersal and Concentration."

conditions created in part by the Hells Gate slides, the industry expanded its spatial focus.

Science and Conservation

The decline in the commercial catch placed new pressures on state regulators to respond to concerns over habitat restoration and to monitor the Hells Gate site for further problems. After 1914, an on-site guardian reported on conditions at Hells Gate for the federal department and Babcock continued his annual inspection. At different times, American authorities, encouraged by an increasingly sceptical Puget Sound cannery lobby, toured the scene and offered more critical assessments, suggesting that problems might still exist at Hells Gate.[82] Pressure on fisheries officials rose as Fraser River canners, starved of their former supply, began to criticize governmental efforts as well. In 1926, two exposés published in the *Province* focused the department's attention by charging that Hells Gate was a menace to salmon and that former clearing efforts had failed.[83]

In a move aimed to stem criticism, restore official legitimacy, and sincerely approach the basis of the concern, a board of engineers was appointed in 1926 under the auspices of the Department of Marine and Fisheries. Was Hells Gate restored to its former condition or could it be improved? Those were the questions put to the board consisting of J. H. McHugh (fisheries engineer, Department of Marine and Fisheries), C. E. Webb (district chief engineer, Dominion Water Power and Reclamation Survey), P. E. Doncaster (district engineer, Public Works Canada), and H. W. Hunt (assistant engineer, Department of Marine and Fisheries). To answer the criticisms, the engineers conducted tests in hydraulics. They pursued a series of investigations at the Hells Gate site on stream flow,

82. NAC, RG 23, Vol. 679, File 713-2-2[12], Arthur S. Einarsen, Washington Division of Fisheries, "A Report on an Inspection Trip to the Fraser River Watershed," June 29, 1929.
83. NAC, RG 23, Vol. 679, File 713-2-2[8], J. McHugh, "Interim Report of the Engineers Enquiring into the Fraser River Conditions at Hell's Gate and Bridge River Canyon," nd. McHugh mentions the *Province* articles in spurring on the first meeting of the Board. The dates of the articles are listed in this document as August 15 and September 4, 1926. I have been unable to locate the original newspaper articles.

velocity levels, and turbulence over different points of the year; they constructed contour maps, painted a new gauge on the Gate's walls, and managed to develop a more finely tuned model of water movement than had existed previously. They found that at low water the surface width of the channel was 85 ft, but could attain a width of 180 ft at high stages. The remarkable variability of the Gate was etched in sharp relief.[84]

The engineers also sought the aid of fisheries officials and recorded their views. Thomas Scott, the federal guardian at the site, believed fish did not encounter unusual obstacles at Hells Gate. As the direct observer of the site since 1913, his views carried considerable weight. Furthermore, Babcock, the provincial assistant commissioner, concurred. He believed that previous work was satisfactory and that further intervention might harm rather than improve current conditions. In confidence, he told one canner that he believed that negative reports emanated "from men who are desirous of getting contracts."[85] However, A. P. Halladay, a federal inspector of fisheries, disagreed. He felt that Hells Gate was still a problem and that more work was required. He hoped the engineers could propose solutions.[86] It was left to an anonymous "Observer" in the Vancouver press to reflect that Hells Gate was and remained "a rock of disappointment, a rock of rages, a symbol of narrow schisms in the midst of torrential progress."[87]

The Final Report of the Board of Engineers released in 1928 provided an ambivalent set of answers to public questions. The engineers weighed their evidence and suggested that turbulence was probably the greatest problem for fish. Turbulence was created by the "conflicting currents set up by the great irregularities on both river banks as well as on the stream bed." To reduce turbulence,

---

84. NAC, RG 23, Vol. 679, File 713-2-2[9], C. A. Webb, "Interim Report on Hydraulic Investigations, Carried Out by Dominion Water Power and Reclamation Service on Hell's Gate, July 1927" (dated July 9, 1927).
85. BCARS, GR 435, BC Department of Fisheries, Box 108, File 1069, Babcock to Bell-Irving, Anglo-British Columbia Packing Company, May 22, 1928 (copy).
86. McHugh, "Interim Report" and NAC RG 23, Vol. 679, File 713-2-2[9], J. A. Motherwell, chief inspector of fisheries, to W. A. Found, director of fisheries, Department of Marine and Fisheries, March 15, 1927.
87. This clipping was found in the papers of the federal Department of Marine and Fisheries: NAC, RG 23, Vol. 679, File 713-2-2[9], The Observer, the *Province*, November 26, 1927.

they recommended straightening the channel to remove the roughness of the edge and pointed to some large jutting rock on the left bank as the key site for reconstruction.[88] Although it is impossible to know, it would be interesting to discover whether this rock-removal recommendation bore any similarities to that suggested by members of the Boothroyd band in 1916. At any rate, conscious of the limits of their knowledge, the engineers suggested a cautious response.

The stumbling block of the report was that it could not answer enough questions for the Department of Marine and Fisheries about how all of this would affect fish. The omission of trained fishery experts from the investigation is striking, and it points to the inability of the department to raise a team competent for the task at this date – despite the existence of a marine biological station under federal authority at Nanaimo, established in 1908 – and to the implicit assumption that problems of construction and river training should involve engineers, and only engineers. At one point W. A. Clemens of the Nanaimo station suggested that tagging experiments should be conducted to measure the rate of fish passage, but his idea was not taken up either by the Board of Engineers or the station at Nanaimo.[89] Faced with a report on physical conditions, senior fisheries officials asked biological questions. "Unfortunately," wrote J. A. Motherwell, the federal chief inspector in the province, in response to the report, "there is evidently no one who can say just what a salmon is capable of doing under the several phases of fluctuating conditions experienced at points where rapid or broken water occurs."[90] Consequently, Motherwell judged, the report failed on the crucial lack of knowledge of salmon and therefore could not be implemented without serious risk of harming fish. No changes were made. Members of the fishing industry, fully briefed on the investigation and the department's conclusions, accepted that the state had responded to their request

88. NAC, RG 23, Vol. 679, File 713-2-2[1], J. H. McHugh, C. E. Webb, P. E. Doncaster, R. M. Taylor, and H. W. Hunt, "Final Report of the Engineers Enquiring into Fraser River Conditions at Hell's Gate, 1926–1928," July 27, 1928.
89. NAC, RG 23, Vol. 679, File 713-2-2[9], J. A. Motherwell to W. A. Found, April 19, 1928. Why it was not taken up is unknown.
90. NAC, RG 23, Vol. 679, File 713-2-2[11], J. A. Motherwell to W. A. Found, October 9, 1928.

for a reexamination and thanked them for their efforts.[91] Here was the board's only success: revising the public perception of the department's legitimacy.[92]

The focused scientific investigation of Hells Gate from 1926 to 1928 demonstrated that scientists and fisheries officials were also challenged by the longer-term consequences of the slides. Ten years and more after the slide cleanup of 1913–14, officials were still responding to doubts and attempting to shore up their legitimacy in the face of industry criticism. The flow of meaning through Hells Gate was as much a political problem for the department as a strictly physical question of flow and turbulence. Its response, privileging hydraulic research over the biological and then jettisoning recommendations for the lack of salmon knowledge, suggests the confused understanding of the problem and the perception of the dangers of any attempt to tamper with the flow – in either its physical or its symbolic aspects.

Conclusion

The Hells Gate slides remade the Fraser River, the salmon, and their claimants. Moving rock changed the river's flow, threw up an obstacle to salmon, and transformed the way that water and salmon confronted the Gate. Throughout its history Hells Gate has been a focusing point of ecological and social power. Along its banks complex systems of social regulation emerged in native societies to control access. After the slides, fisheries officials annexed authority at Hells Gate; native resource rights were ignored and overridden. Pressed by a commercial fishery with an expanding appetite for product, fisheries officials attempted to repair the difficulty and later restore confidence in the soundness of conditions.

91. NAC, RG 23, Vol. 679, File 713-2-2[11], "Conference re Hell's Gate Conditions Fraser River," November 21, 1928.
92. Fisheries officials were highly conscious of the public perception of their decisions. When reports of low fish returns gained wide attention in 1927, C. W. Harrison, the federal district inspector of fisheries, advised his superiors not to engage in dip-net assistance as this would merely set a precedent and "it would be acknowledgement that the Department considered the fish were unable to make their way through the canyon." BCARS GR 435, BC Department of Fisheries, Vol. 107, File 1064, C. W. Harrison to Motherwell, September 29, 1927 (copy).

The natural and social changes stemming from the slides were experienced within shifting spatial realms. At Hells Gate, native access was limited; fisheries officials incorporated the canyon as a zone of control and concern; the fishing industry gained influence in the regulation of salmon spawning habitat. By restricting salmon migration, the slides created new natural spatial limits: Above the canyon salmon became more rare and in the ocean less numerous. The salmon's claimants, native peoples and the commercial fishery, redirected the location of their activities: Native peoples shifted their economies toward more hunting, gathering, agricultural, and wage labor activities; the commercial fishery focused on different species and redirected the bulk of fishing effort to northern rivers. A landslide, triggered by a technology aiming to compress time and space, produced a cascading set of effects that reordered the natural and social spaces of BC. Fraser's legacy collided with Coyote's world.

2

Damming the Tributaries

At Hells Gate, British Columbians had attempted to set the river free, but they had more experience in the opposite direction. Since the early days of the gold rush, starting in 1858 on the Fraser, the resettlement of BC had consequences not only for the land and for native space, but also for the rivers and lakes of the Western Cordillera. Small rivers had been diverted during the gold rush to supply placer mining outfits. Farmers had drawn from streams to water cattle and, increasingly, to supply irrigation districts in the semiarid regions of the interior. In the lower Fraser Basin, tributaries had been rerouted, diversion dams thrown up, and lakes drained to make way for more farmland. Burgeoning urban centers had reached into their hinterlands for domestic water supplies and repiped coastal rivers. All of these changes had shifted the social power over rivers and lakes from native to newcomer.[1] From the vantage point of the twentieth century, they were just the beginning. With the expansion of the extractive resource economy and the rise of hydroelectricity after 1900, the rivers of BC came under increasing scrutiny as power sources and development sites. Diversion gave way to dreams of damming.

Before World War II, these dreams focused on tributaries and smaller systems. With the exception of one ill-fated proposal to dam the Fraser Canyon near Hells Gate in 1912, the idea of controlling the main stem of the Fraser was practically unimaginable.[2]

1. Cameron, *Openings*; Harris, *The Resettlement*.
2. CSB, "Will Halt River and Erect Huge Power Plant," *The World*, June 24, 1912; "Salmon Canners May Oppose Dam," *The World*, June 25, 1912; BCARS, GR 435, Department

Instead, power firms dammed lower-basin tributaries that could be developed close to market and within the limits of economy and demand; mining ventures expanded hydraulic and placer mining activities in the canyon and the upper basin; and lumber firms built splash dams to assist log transport. Across the basin, the pattern of development was dispersed, relatively modest, and did not hinder salmon runs as much as if a main-stem development program had ensued.

Neighboring jurisdictions experienced a different development path. Rather than avoiding main-stem river projects, developers across the continent looked increasingly to large rivers as the most advantageous power sources. To the east, in 1911 power interests dammed the Bow River in Alberta to electrify Calgary. To the south, in the U.S. Pacific Northwest, federal agencies began to raise large main-stem dams on the Columbia River in the early 1930s to supply irrigation projects, as well as to provide navigation benefits, flood control, and hydroelectric power. Across North America, from the Niagara River in Ontario to the Colorado River in the U.S. Southwest, entrepreneurs, corporations, and government agencies seized development opportunities and dammed main stems. By the 1930s, promoters announced the arrival of the big dam era, but not in BC.

In some respects, this outcome was surprising. The province, after all, had experienced strong economic growth from 1900 to 1930, based on the export of key staple resources, including timber, minerals, agriculture, and fish. Vancouver had emerged as the primary Pacific metropolis in Canada, connected by transportation and communications linkages to international markets, transcontinental trade, and a diverse resource hinterland. BC's expanding resource economy sat astride a province richly endowed with rivers and lakes that could be converted to waterpower. Since the turn of the century, international investment capital had begun to develop BC rivers, particularly in the Fraser's lower basin. With grander ambitions, the provincial government had begun to promote

of Fisheries, Box 61, File 565, Daniel MacDuff to Babcock, December 11, 1912; "Information for the Department of Fisheries, Victoria, BC in Connection with the Application of the International Railway and Development Company, Ltd, Vancouver BC for Water Power Rights on the Fraser River" (nd); File 570, deputy commissioner of fisheries to J. F. Armstrong, acting comptroller of water rights, December 5, 1912 (copy).

development opportunities by surveying rivers across the province and publicizing them to international audiences.

In 1919 a compendium of hydrological research published by the Canadian government sought to explain why the Fraser would not be developed soon. First, there was the problem of salmon. The fishing industry was large and contributed a substantial sum to provincial revenues. The Hells Gate episode, it underlined, had proven only too clearly the dangers of obstructing the river. Second, railroads followed the river through the Fraser Canyon and any storage reservoir would risk flooding them. Third, there were smaller tributaries in the lower basin that offered power-development sites within a short distance from the major urban market in Vancouver. Fourth, on the lower river – again, near Vancouver – there were no suitable storage sites to build a reservoir. The river, the publication seemed to suggest, was not designed in such a way that it could be used to maximum hydroelectrical effect given the existing settlement pattern. Had Vancouver sat in the Fraser Canyon, had no lower-basin tributaries existed, and had the salmon disappeared after the Hells Gate slides, then, perhaps, the course of development would be different. But within the imaginable limits of transmission, market demand, and physical geography, federal officials proclaimed that the Fraser would run freely for many years to come.[3]

This was an interesting hypothesis. It emphasized rational planning, weighing costs and benefits, and considering consequences. As it turned out, the Fraser's future would be shaped by a set of forces more distant and abstracted from the river. The Fraser would be calculated by engineers and surveyors as a statistical flow figure and as a cost of development; its tributaries would be shaped by corporate strategies focused on consumption patterns in the city and the cut and thrust of competition; the effects of development on salmon would gain attention, from time to time, but without much focus or foresight. This river would not be planned. Rather, it would be affected incidentally by decisions made about related matters. The meanings and purposes of the river would take on a new and transformed light against the promise and glow of electrification.

3. White, *Water Powers*, p. 231.

Hydroelectricity: The First Stage

The arrival of electricity as a widely used public good occurred gradually across North America in the late nineteenth century. Spectacular arc-lighting displays at public events promised daylight at night, and streetcars connected to overhead wires began to move workers in urban centers. In this first stage of development, electrification remained an urban phenomenon. BC followed the general pattern.

Urban demand drove the emergence of hydroelectricity in the Fraser Basin (see Map 4). Although mining activity in the provincial Southeast and forestry development on the coast initiated a number of dispersed, industrially oriented systems, no large resource-related hydroelectric ventures appeared in the middle or upper Fraser Basin before the mid-1930s.[4] Electricity, when used, derived primarily from small, thermal-powered plants. In Vancouver, by contrast, a series of integrated water-development projects were developed in the opening decades of the twentieth century to supply growing transportation, industrial, and domestic electricity markets. These early projects represented a relatively modest water-development program, shaped by development opportunities and costs, the perceived limits of market demand, and the cautious investment and corporate strategy of the province's key electrical utility.

Vancouver grew from lumber town to Pacific railway terminus in the 1880s and quickly gained the electrical accoutrements of larger centers. Small electrical systems, powered by steam engines, appeared by the late 1880s and early 1890s.[5] They energized a skeletal system of streetcars and lights, organized and financed by local businessmen who were supplied through continental networks of technological transfer. In their first decade, these original systems faced declining profits and volatile demand. By 1895, a group of British investors with organizational expertise in urban utilities and access to development capital consolidated these systems into one corporation, the British Columbia Electric Railway Company

---

4. On electrical development in the Southeast, see Mouat, *The Business of Power.*
5. Armstrong and Nelles, *Monopoly's Moment,* Chapters 2, 4; Green, "Some Pioneers"; Nye, *Electrifying America,* Chapters 1–3; Roy, "The Illumination of Victoria."

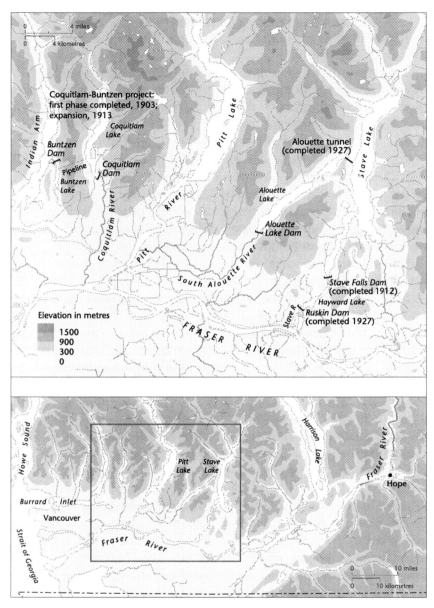

Map 4. Lower-basin dam projects.

(BCER). Focusing on an emerging market with possibilities for consolidation and expansion, the company's directors moved quickly to secure their position.

Before committing themselves to a development program in BC, BCER directors ensured that their investments would be protected from state competition. In various markets across North America and Europe, utilities had faced, and would yet face, the unwelcome pressure of civic populist groups and the threat of government takeover following an initial stage of market development. Rather than confront such problems down the road, the BCER called on the province and city of Vancouver to grant special rights to consolidate preexisting firms, remove the possibility of municipal competition, and open new utilities fields, such as telephones, in the future.[6] The city and the province complied. A consolidated-utilities scene offered stability and permanence to a previously fractured and undercapitalized industry. At this date, the city of Vancouver had no intentions of entering the market as a competitor and the agreement secured a major investor in the province. To maintain its strong position, the BCER also cultivated relations with the long-serving provincial Conservative government under Premier Richard McBride and reminded him when necessary of the power of the BCER to disrupt the province's credit rating on the London markets should regulation prove too onerous.[7] With a supply monopoly, governmental acquiescence, and a long-term position established in the market, the BCER moved to upgrade Vancouver's electrical system. The water flowing in the city's hinterland would soon contribute to the pulse of urban life.

Surveyors looked inland from the city of Vancouver and asked where power might lie. When BCER river surveys began in 1898 under the subsidiary, Vancouver Power Company, there were no state-funded hydrological maps or compendia available. There were, in fact, no official registers of provincial river flows. Only after 1911 would the federal Commission of Conservation initiate a stream-gauging program in BC, mimicking the U.S. Geological Survey's national inventory.[8] In the meantime, surveyors had to

6. Armstrong and Nelles, *Monopoly's Moment*, pp. 97–8.
7. Roy, "The Fine Arts of Lobbying," pp. 244–5.
8. Girard, *L'ecologisme retrouvé*, p. 121; White, *Water Powers*, p. 3.

begin where they could. They scattered across the urban hinterland, looking at tributaries, assessing storage possibilities, and staking water rights. To gain a sense of annual stream flows, they asked local inhabitants whether rivers seemed high or low and calculated estimates of "normal" flow accordingly.[9]

From the beginning, surveyors ignored the Fraser River, the largest potential source of waterpower in the provincial southwest. With a growing but still limited local market, the economic justifications for a major river project were not apparent. British investors backing the BCER hoped not for a massive and speculative capital investment, but for orderly and sustained growth. Thinking of the river, surveyors saw that dam sites near Vancouver did not exist. From Hope to the sea, the river descended at a gentle decline, depositing rocks and gravels through the Fraser Valley before diverging into branches through the delta. There were no obvious canyons or drops in this stretch of the river where a dam could be placed and no obvious sites for reservoirs. A major river development would have to be located further inland, where the sharp declines and sheer narrows of the Fraser Canyon offered possibilities. This would present a significant challenge to existing transmission technologies and add enormously to development and operating costs.

The surveyors turned their attention instead to a series of small lakes, feeding tributaries that joined the Fraser in its final reach to the sea. One by one, these lakes and rivers were measured and gauged. Surveyors and engineers looked for dam sites with storage reservoirs in lakes and calculated the costs of wiring small power plants to transmission lines linked to the city beyond. The Coquitlam River, among a number of sites, proved the most attractive to BCER surveyors. The river was the closest of the available choices to the city, and a potential diversion scheme promised a substantial power source. In any event, the firm had been beaten to another

9. UBC, BC Electric Railway Company (hereafter BCER) Papers, Box 22, File 521, Charles A. Lee, assistant engineer, to G. R. G. Conway, chief engineer, December 7, 1911; Lee, "Report on the Power Resources of the Coquahalla River." For similar descriptions of survey research practice, see File B1384, "Extract From a Report by Sanderson and Porter, March 31, 1908," which discusses surveys on the Cowichan River; and Box 121, File 7, James T. Garden to F. S. Barnard, manager of the Consolidated Mining Company, April 7, 1897, describing the Stave Falls.

potential river, the Stave, by a group of local businessmen who had surveyed and gained rights to the site in 1897 and organized the Stave Lake Power Company in 1901. The firm controlled the site but did little with it initially.[10] In 1902, unfazed by the potential of competition, the BCER proceeded to develop the Coquitlam River.

The project dammed the Coquitlam River, reversed its flow, and joined two formerly separate lakes to make one source of power. In 1903, as the Coquitlam River backed up in Coquitlam Lake behind a 19-ft dam, pipelines, stretching 2.25 miles in length, diverted the stored water out of the watershed and into Buntzen Lake, which became a reservoir. Four-hundred feet below, the company built a small powerhouse on the north arm of Burrard Inlet to capture the surging flow, pass it through turbines and send the converted 1500 kW of electrical energy on transmission wires to Vancouver.[11] When it opened in 1903, the project not only marked a technological advance on the west coast of Canada, but also demonstrated the rise of Vancouver as a regional metropolis, with the infrastructure to match its ambitions and sense of self-importance. At its inauguration, a group of Vancouver businessmen, attired in black formal wear and bowler hats, stood at the base of the pipelines and welcomed the new day of power as it surged around them (see Photograph 5).

Shifting from steam power, driven by relatively expensive sources of coal to hydropower that promised to be substantially cheaper, the BCER could risk dropping its rates to test the limits of local demand. In part this business strategy aimed to fend off the challenges posed by the Vancouver Gas Company, which the BCER quickly purchased in any event. But it was also a marketing strategy proven in urban markets across Europe and North America: Drop rates and demand will climb.[12] In the years after the completion of the Coquitlam project, new markets emerged. From 1906 to 1912, the

10. Armstrong and Nelles, *Monopoly's Moment*, p. 100; Roy, "The British Columbia Electric Railway Company," p. 157.
11. Pacific Salmon Commission Library and Archives (hereafter PSCA), File 1180 1–15 "Obstructions – History – Dams – Lower Fraser," p. 1; UBC, BCER clippings file (hereafter BCER CF), "Vancouver's Power Sources Cost Many Million Dollars," *Vancouver Sun*, May 30, 1927; White, *Water Powers*, p. 151.
12. Armstrong and Nelles, *Monopoly's Moment*, p. 99.

Photograph 5. Opening of the tunnel connecting Coquitlam and Buntzen Lakes.
(*Source*: Arthur V. Davis, *Water Powers of Canada*, Ottawa: Commission of Conservation, 1919.)

BCER's list of commercial customers rose from 246 to 2555.[13] An expansion to the project beginning in 1909 and completed in 1912 increased the size of the original diversion dam and pipelines and expanded the capacity of the powerhouse.

In the background to the BCER's Coquitlam expansion, investors reorganized the dormant Stave Lake Company as the Western Canada Power Company (WCPC) and proceeded with a competing venture on the outlet of Stave Lake. The lake sat about 35 miles east of Vancouver, north of the Fraser River, into which it drained. The Stave site promised great potential power possibilities through the combination of storage in the lake and a substantial head, created by an 80-ft-high set of falls.[14] In the language of contemporary journalists, the project was the natural counterweight to the Coquitlam–Buntzen project in the local geography of power, as if fashioned by an "All Wise Providence."[15] The lake provided a ready reservoir site, and the company would simply raise a dam at its outlet to capture the force of gravity in the falling water. The Stave Lake development reached completion in 1912, the same year the BCER finished its Coquitlam expansion.

Although the rivals might have run up demand in the local market through a competitive price war, they opted instead for a safer strategy: collusion. In a growing market, the two corporations struck an agreement to divide the customer base. The BCER held on to lighting and heating contracts, as well as those carrying a load of less than 120 kW, whereas the WCPC gained industrial customers carrying a load of over 150 kW. Contracts in between these two categories remained open to competition.[16] Added to the marketing agreement, the BCER promised to purchase from the WCPC at least 5000 kW annually and expand that figure over twenty years. With increasing demand on the horizon, this agreement provided the BCER with a flexible supply.[17] Eventually, in 1920, the BCER would purchase the WCPC outright. In the meantime, it enjoyed

13. Roy, "The British Columbia Electric Railway Company," p. 175.
14. An original letter of appraisal of the site for the Consolidated Mining Company is contained in: UBC, BCER Papers, Box 121, File 7, James T. Garden to F. S. Barnard, Consolidated Mining Company, April 7, 1897.
15. BCER CF, *Columbian*, August 13, 1910.
16. Roy, "The British Columbia Electric Railway Company," p. 174.   17. Ibid.

the benefits of an expanded supply and a protected market niche without bearing any of the financial risks of development at the Stave site.[18]

Economic growth in the 1920s transformed Vancouver's electricity market. Following a recession after World War I, the city experienced rapid growth in manufacturing and resource export industries. The opening of the Panama Canal helped to solidify Vancouver's position as Canada's Pacific metropolis; it facilitated the export of a host of BC products from apples to zinc and redirected trade to and from the Prairies through the port. BC's manufacturing increased from a net production value of $28 million in 1922 to $134 million in 1929.[19] In Vancouver the number of manufacturing concerns grew over the decade by over 50% – from 441 in 1921 to 681 in 1931. Burgeoning trade and manufacturing attracted migrants to the city; the population doubled in ten years to 246,593 by 1931.[20]

Growth bore consequences for the BCER system. Transmission lines fanned out across and beyond the city to reach new neighborhoods and carry electricity to neighboring municipalities. These lines electrified a growing consumer base. Across the province, the number of domestic power customers almost doubled from 69,909 in 1920 to 125,171 in 1930.[21] The corporation responded haltingly to the new consumer market, promoting some domestic goods, like irons, to sell electricity in off-peak hours, but moving slowly to upgrade infrastructure and extend transmission to municipalities at a distance. Higher returns could be generated from the new manufacturing market; smaller investments in transmission lines to key sections of the city led to a burgeoning industrial market. Lumber processors that had previously relied on burning wood waste for power increasingly engaged the BCER, and expanding industries like oil and sugar refineries added to the demand. Above all of these new demands, however, remained the core business of urban transportation. In lockstep with economic and population growth, the BCER streetcar system expanded its routes in and around the city

---

18. City of Vancouver Archives, Ad MSS 321 BC Electric Co. Ltd., Vol. 1, File 1, "BC Electric Railway Company General History," p 18.
19. Barman, *The West Beyond the West*, pp. 237–40.   20. Roy, *Vancouver*, p. 168.
21. Mary Doreen Taylor, "Development of the Electricity Industry," p. 45.

and carried evermore customers. The streetcars accounted for the single greatest draw on BCER electricity during the 1920s.[22] Historians of electricity have noted the remarkable growth in electrical demand in North America during the 1920s as electrification came to be seen as a necessity of modern life and as a key component of urban expansion and economic advantage.[23] Vancouver was no exception.

The BCER met the challenge of the growing market of the 1920s by expanding its supply network. The company fended off renewed regulatory attempts at the provincial and city levels and cannibalized rivals and small concerns as it launched into an ambitious building program.[24] It purchased the WCPC in 1920 and conducted a substantial refit of the Stave Lake dam and its powerhouse in 1924. To increase the potential flow of the Stave project and its generation potential, the BCER diverted neighboring Alouette Lake, damming its outlet on the Alouette River and piping its flow into the headwaters of Stave Lake. The development followed the pattern earlier established in the Coquitlam–Buntzen project, capturing the flow of parallel watersheds into a single focus. The company also reached beyond the urban core region. In 1925, the BCER considered the potential of sites in and around the Fraser Canyon and bought the rights to a power site on the Bridge River, earlier developed by a small mining company. Farther in the interior, the BCER purchased the electrical utility in the city of Kamloops, along with its power source, a small hydroelectric facility on the Barriere River, a tributary of the Thompson (see Figure 1).[25]

By the late 1920s, as urban expansion in the Vancouver region continued apace and the province's natural resource export markets boomed, it seemed as if the growth would never end. In 1927, a BCER engineer estimated that the firm would experience an annual increase of 10% in its total electrical load, which would require a doubling of output in seven or eight years. Mainland plants produced a total of 332,000,000 kWh of power in 1925. To

22. Roy, "British Columbia Electric Railway Company," pp. 305–6.
23. Brigham, *Empowering the West*, p. 8.
24. Armstrong and Nelles, *Monopoly's Moment*, pp. 261–2.   25. Wittner, "Barriere."

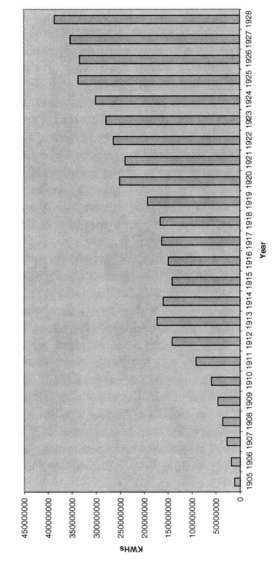

Figure 1. BCER mainland power production (in kilowatt hours), 1905–28.
(*Source:* Roy, "The British Columbia Electric Railway Company.")

65

double this capacity by 1934, around $25 million would need to be invested in new hydroelectric projects, transmission, and related facilities.[26] The BCER raced to meet production targets to bolster its supply position. Company officials considered two large new projects at Bridge River and on the lower Stave. The first promised a considerable increase in potential power, despite a relatively long transmission distance from the Fraser Canyon.[27] The second was a smaller project, but it held the advantage of tying into existing BCER facilities: It would be sited at Ruskin below the existing Stave Lake dam, putting falling water through another turnstile, so to speak. Further, its power could be delivered through established transmission lines. Ultimately both projects would be announced, but only the Ruskin project would move forward in 1928. The growth of power demand at the time required an almost immediate addition to the BCER's power supply, and the Ruskin project could be expedited at a lower cost than that of the Bridge River alternative.[28]

BCER managers still worried that the firm might not be able to keep pace with demand. W. G. Murrin wrote internal memos to his subordinates, fretting about a power shortfall by 1929: "Unless the Ruskin plant is in operation by October of next year, Vancouver will be short of power and her industries on part time, and the time schedule is a very close one and does not allow for any unseen delays."[29] The problem, it turned out, was worse than that. Precipitation in the lower basin in the summer and fall of 1929 was lower than average.[30] The level in BCER reservoirs fell accordingly, affecting the potential power supply. Nature's vagaries had combined with increased market demand to make the BCER's margins thin indeed. The looming shortfall was averted only by the rapid construction of the Ruskin dam. Four hundred and seventy-nine

26. Carpenter, "Water Developments."
27. BCER CF, "Choose Route of Power Line," *Province*, May 26, 1928.
28. UBC, BCER Papers, Box 158, File 6, A. C. Clogher, hydraulic engineer, to C. E. Calder, president, American and Foreign Power Co., NY, January 16, 1929 (copy); Murrin diary, December 15, 1928, "Interview with Clogher."
29. UBC, BCER Papers, Box 158, File 6, Murrin to Goward, May 3, 1929 (copy).
30. UBC, BCER Papers, Box 158, File 6, Murrin to N. S. Braden, November 25, 1929 (copy).

men installed the powerhouse and dam in a year at a cost in the range of $6 million.[31]

The BCER's ambitious building program came to completion in 1929 just as the economic conditions shaping its industrial and domestic market were about to change for the worse. As a staples-driven resource economy heavily dependent on export markets for its products, BC fared poorly when international economic shocks in the early 1930s reduced international demand and produced barriers to trade, particularly in the United States. In the late 1920s Canada had ranked among the most export-oriented industrial economies in the world. Kenneth Norrie and Douglas Owram state that, whereas 22% of Canada's national expenditure derived from merchandise exports in the late 1920s, only 5% did in the United States.[32] Much of this export trade depended on a few core trade items; farm, animal, and forest products accounted for nearly 80% of all Canadian exports. Canada's export economy was not only large; it was also narrowly based on the American market. In 1929 one-third of Canadian exports went to the United States.[33] British Columbia's economy presented some differences from the national pattern – fewer agricultural exports, more forest and fish products – but it was just as vulnerable. When the prices of and demand for Canada's export commodities declined in international markets in the early 1930s and when the United States imposed barriers to trade in 1930 under the Hawley–Smoot tariff, the narrow and thin basis of Canadian prosperity deteriorated.[34]

As a result, the outlook of the power industry in BC changed. Contracting economic activity in the primary resources sector led to a decline across the electrical-demand curve. Over the 1930s the company accommodated limited rises in consumption by simply

31. UBC, BCER Papers, Box 158, File 6, Mr. Carpenter to Murrin, July 24, 1929. Another report, dating from 1940, pegged the cost of the first Ruskin development at $6,470,468.65; same file, E. E. Carpenter, Construction Department, to Murrin and Adams, March 6, 1940.
32. Kenneth Norrie and Douglas Owram, *A History of the Canadian Economy* (Toronto: Harcourt Brace Jovanovitch, 1991), p. 480.
33. Ibid.   34. Ibid., pp. 475–89.

installing more powerful generators at the Stave Lake site.[35] In other parts of North America, public and private utilities took advantage of low labor and materials costs in the 1930s to build new power-generation facilities and lead market demand with low-price structures. BCER directors, to the contrary, decided not to boost the market with cheap power, but followed demand as closely as they dared, the better to save capital. When World War II arrived, this Depression-era strategy would come back to haunt the firm.

The inauguration of the Ruskin dam in November 1930 held up a mirror to BC and revealed the aspirations and tensions that the promise of power contained. The official ceremony juxtaposed social power and turbines in a garish moment of self-congratulation. Two-hundred local businessmen and politicians – including the premier and the lieutenant governor – dined in the powerhouse, toasting the progress of power, while aware, nevertheless, of the tenuous state of the BC economy.[36] Across the city, newspapers printed special editions, and newsreels of the dam played in cinemas.[37] The new Ruskin power was announced as if a sign of change. In the following years, visitors flocked to the dam to witness modern technology transforming white water into power. Between 1932 and 1937, the BCER recorded 30,000 visitors taking tours at the Ruskin facility – this on the outskirts of a city of some 250,000 inhabitants.[38] Perhaps as never before, the Ruskin dam provided a needed and desired symbol of expansion. Read retrospectively, however, the dam could be only an ironic symbol; its rise marked the end of Vancouver's power boom, not its continuation. The decade to come would witness no new building in the Fraser Basin.

From an early investment stage to the 1930s, the BCER's corporate strategy had operated according to two principles, caution in investment and concentration of development. An early dominance

---

35. PSCA, File 1180 15, "Obstructions – History – Dams – Lower Fraser," April 18, 1940, p. 4.
36. UBC, BCER Papers, Box 158, File 6, Murrin to Premier Tolmie, October 23, 1930 (copy); Tolmie to Murrin, October 29, 1930; Murrin to Fairburn, October 24, 1930 (copy); A. M. D. Fairbairn, executive secretary to lieutenant governor, to Murrin, October 21, 1930.
37. UBC, BCER Papers, Box 158, File 6, Murrin to Howard, December 4, 1930 (copy)
38. UBC, BCER Papers, Box 158, File 6, A. Vilstrup to Murrin, February 5, 1937.

in the market and a careful treatment of competitors had allowed the company to avoid risky investments and focus on meeting predictable increases in demand. This position had allowed the firm to concentrate its efforts on a number of key sites, maximizing power yielded from watersheds through intensification strategies rather than pursuing extensive development on other lakes and rivers. For a set of reasons utterly unrelated to environmental or fisheries protection, the early stage of hydroelectricity in the Fraser Basin had avoided some of the more egregious potential environmental effects of damming witnessed in other North American regions in this period. The BCER had produced an illuminated, electrified urban system on the Pacific, but it had completed these tasks without damming the Fraser.

The Fisheries Question

The BCER and its competitors did not act only in a business environment. Other interests raised questions about how, when, and where rivers should be dammed. Governments sought to impose limits on the form and scale of development. The rivers of the Fraser Basin were not infinitely malleable; they represented contested space. But to what extent did the fisheries question affect dam development?

Before dams rose in Vancouver's hinterland, a resource extraction dam on the Quesnel River in the Fraser's upper basin suggested some of the problems that dam development posed to fisheries conservation and to the practicalities of government regulation. Built in the late 1890s by mining investors, the dam diverted the river in order that its bed might be mined by use of placer techniques, from the example of similar ventures in California.[39] Located at the outlet of Quesnel Lake, the dam measured 18 ft high and had a length, from end to end, of 763 ft. Built of wood and rock, the dam

---

39. PSCA, File 1180.1–17, "Quesnel Dam," McNab to deputy minister of marine and fisheries, May 4, 1899. On the problem of mining and rivers in California in the nineteenth century, see McEvoy, *The Fisherman's Problem*, pp. 83–4; Merchant, *Green Versus Gold*, Chapter 4, "The Environmental Impacts of the Gold Rush," pp. 101–39; Mount, *California Rivers and Streams*, pp. 202–6.

wrapped around the lake in an arc with a radius of 460 ft. It was relatively large, had few openings besides a raceway and a small fishway, and blocked a system of lakes and rivers. Beyond Quesnel Lake, the Horsefly River had produced major sockeye runs on the big-year cycle, but beginning in 1899, these runs were effectively blocked.[40] "No other condition affecting the spawning grounds of the province is of such pressing moment," John Pease Babcock stated in 1902, in one of his first reports on the condition of BC fisheries.[41]

Protest against the dam long preceded Babcock's observation. At the time of the dam's completion, the Fraser River Canner's Association wrote to the federal Department of Marine and Fisheries to protest the threat it posed to the spawning grounds. "This dam," wrote Association Secretary W. D. Burdis, "without doubt, cuts off *many thousands of miles of the best spawning grounds* in the province."[42] MP Hewitt Bostock told the minister of marine and fisheries in 1899 that he had visited the dam that season and had found evidence of a serious problem. An engineer at the dam said that fish had died in enormous numbers while struggling to pass upstream. So many salmon had expired at the site that workers had placed them in great "heaps" on the shore and burnt them.[43]

Damaging though the dam was, no public agency seemed ready or willing to bring its owners to heel. No federal or provincial water officials expressed interest or concern. The federal Department of Marine and Fisheries delayed and may or may not have demanded a fishway, depending on different reports.[44] The fishway that existed at the dam site did not work properly and could pass only a small proportion of the annual runs. Federal fisheries regulation, at this date, operated at a distance, with fragmentary knowledge of the case and little understanding of the migratory nature of Pacific salmon.

40. Babcock, *Report of the Fisheries Commissioner of British Columbia*, p. 11; Bocking, *Mighty River*, p. 81.
41. Babcock, *Report of the Fisheries Commissioner*, p. 11.
42. UBC Koerner Microfilms, NAC, RG 23, File 2235, part 1, W. D. Burdis to deputy minister of fisheries, May 20, 1899.
43. Ibid., Hewitt Bostock, MP, to Louis Davies, minister of fisheries, June 5, 1899.
44. PSCA, File 1180.1–17, "Quesnel Dam," W. D. Hardie to Gordeau, deputy minister of fisheries, May 20, 1899; CSB, *Province*, May 17, 1901; Babcock, *Report of the Fisheries Commissioner*, p. 13; PSCA, File 1180.1–17, "Quesnel Dam," R. Prefontaine to R. F. Green, November 20, 1903 (copy).

The first serious government effort to alter the dam came in February 1904. A group of canners, representing all the major firms operating on the Fraser, arranged a conference in Victoria with newly elected Premier Richard McBride to press for action on the dam problem.[45] Although the question of regulating dams for the fishery was a federal responsibility, the canners had tried unsuccessfully to influence Ottawa in 1898 and 1899. They wished for the provincial Fisheries Commission to step into the void and establish an operational fishway. Apparently without concern for jurisdictional complications, McBride approved the idea. Shortly after the meeting, the canners sent a telegram to a Mr. Hobson, a representative of the dam's current owner, the Cariboo Hydraulic Mining Company, instructing him to wire approval for the project to the provincial government.[46] The consent was apparently received, for by the time of the 1904 spawning season Babcock had installed a much wider fishway that, by all reports, was a great improvement. Using the provincial Fisheries Commission – even corresponding on its behalf – the cannery interest had forced provincial action and bypassed federal jurisdiction entirely. The federal department, on the other hand, appeared, not pliant, but inattentive.

The debacle of the Quesnel Lake case may have had some limited effect in focusing federal attention on the problem of dams. In 1909, following reports of problems with a dam on the Adams River, used for flushing logs downriver to the town of Chase, federal officials took an active role in cooperating with provincial Deputy Commissioner Babcock to establish a satisfactory fish passage. Although the fishway proved to be deficient, according to later fisheries scientists, there was, at least, an attempt made to force the company to comply with the federal Fisheries Act. The failure in this case was not due to federal inattention in the first instance, but to the rudimentary state of fishway technology and the lack of continued observation.[47] As in the Quesnel case, the hinterland

45. CSB, *Vancouver News Advertiser*, September 9, 1903; *Colonist*, September 29, 1903; UBC Special Collections and Archives, IPSFC Collection, *Minutes of Committee Meetings of the Fraser River Canners' Association, Vancouver, BC (1900–1904)*, p. 294, re: meeting, February 4, 1904.

46. UBC, IPSFC Collection, *Minutes of Committee Meetings of the Fraser River Canners' Association, Vancouver, BC (1900–04)*, p. 294.

47. PSCA, File 1180.1–14, "Adams River Dam," inspector of fisheries to Department of Marine and Fisheries, February 24, 1909 (copy); inspector of fisheries to W. A. Found, September 26, 1911 (copy); chief inspector of fisheries to W. F. Richardson, September

location had a detrimental effect on the ability of fisheries officials to assess problems on an ongoing basis.

The Quesnel dam was finally dismantled in 1921.[48] The dream of gold had faded, and the danger of a dam break provided the excuse for its wholesale removal, but not before the dam had done considerable damage to important upper-basin spawning grounds. In its early years, depending on water levels at the entrance to the original fishway, the dam played havoc with runs. Although runs allegedly passed the raceway of the dam in 1898 and up the fishway in 1901, no salmon passed in 1899; in 1900 the runs were delayed three weeks; and in 1902 they stayed below the dam for nine weeks, many dying unspawned.[49] The success of spawners after the installation of a new fishway in 1904 was much better according to observers, but how much better is difficult to say. The dam still blocked passage to spawning areas in Quesnel Lake and the Horsefly River that provided the largest single site of spawning habitat in the basin.[50] Retrospective reconstructions of salmon populations suggest that, along with overfishing, the Quesnel dam accounted for major swings in salmon populations between 1898 and 1904.[51] After that time, the populations did rebound, only to be hindered once again when the slides in the Fraser Canyon affected all upper-basin areas. The Quesnel Lake dam had a far greater impact on Fraser River salmon than did any other dam in the basin before the 1940s.[52]

In the lower basin, dam development enjoyed high praise from civic leaders, but also attracted a diverse set of critics. The BCER's plan to dam the Coquitlam River in 1903, for example, created conflict. On the lower Coquitlam River, the Coquitlam band, a Coast Salish group, held a small reserve on both sides of the river.

12, 1911 (copy); inspector of fisheries to Department of Marine and Fisheries, June 27, 1910 (copy).

48. NAC, RG 23, Vol. 829, File 719-9-26[1], "The Removal of an Historical Landmark," *Canadian Fisherman*, January 29, 1919; CSB, "Great Dam is Destroyed at Quesnel," *Province*, June 9, 1921.
49. PSCA, File 1180.1–87a, A. C. Cooper, "The Causes of Decline of the Quesnel River Sockeye Runs," 1951.
50. Babcock, *Report of the Fisheries Commissioner*, p. 11.
51. Thompson, *Effect of the Obstruction.*
52. CSB, "Streams Once Fairly Choked with Salmon," *World*, July 11, 1917; Hume, *Adam's River*, pp. 83–4; Roos, *Restoring Fraser River Salmon*, pp. 16–17; Thompson, *Effect of the Obstructions.*

Fishing sites existed on the reserve and were used to catch coho, chum, and sockeye salmon that spawned in the Coquitlam River and Lake. When rumors of a dam first circulated in 1899, probably around the time of surveys, Chief Johnnie of the Coquitlam band wrote to federal fisheries officials and outlined the threat posed to his band's way of life[53]:

I heard that the Vancouver people were going to buy Coquitlam creek ... I heard that they want to dam it and turn it into pipes that will stop the salmon from hatching here. We are all loyal subjects of the Queen and would like to be given a chance to live honestly and comfortably[.] [I]f the creek is taken away from us ... it is like a man taken the food out of my cupboard.

Chief Johnnie went on to describe the river as the "storehouse" of his people and asked for annual compensation for thirty people, should development proceed. The damming of the Coquitlam River threatened to damage the long-standing fishery of the Coquitlam band and reshape the character of their relationship with the river. Pipes, not fishnets and weirs, would come to control and reorganize this space, replacing detailed local knowledge of a fishery and its changing character. Perhaps unsurprisingly, there is no evidence that Chief Johnnie's concerns played a role in the planning of the Coquitlam–Buntzen project, nor that compensation was paid.

The BCER and fisheries officials were more concerned about the reaction of organized commercial fishing interests than native protests. As in the Quesnel case, the Fraser River Canners Association raised a number of concerns about the effects of development on local salmon runs, and they succeeded in forcing some remedial measures in the initial design. Guided by the federal Department of Marine and Fisheries, the BCER duly installed a 12-ft-wide fishway in the Coquitlam dam and agreed to release water over the dam in May when the sockeye ran.[54] Despite these actions, complained the Fraser River Canners in 1906, the fishway did not work. W. D. Burdis, secretary of the canners association, had information from

53. UBC Library microfilm reel 52 of NAC RG 23, File 2780, part 1, Chief Johnnie to? stamped by inspector of fisheries, New Westminster, March 19, 1899, signed with an X.
54. PSCA, File 1180.1–12, "Coquitlam Dam," C. B. Sword to E. E. Prince, Dominion commissioner of fisheries, April 2, 1904.

a local observer in 1905 that not 1 of over 1000 migrants reaching the base of the dam passed through the fishway. Part of the problem was surely technical, but Burdis offered a more conspiratorial possibility: The firm was deliberately ignoring the fishway in order to fend off another set of critics entirely – the local municipalities.[55]

Before becoming a power reservoir, Coquitlam Lake was not only a salmon spawning habitat but also a domestic water source for the municipality of New Westminster. After the dam's construction, municipal politicians frequently raised the fear that the water supply would be made impure. Rotting salmon, no longer cleared by river circulation after spawning, posed one such possible source of pollution. The general flooding produced other problems as well. Preexisting vegetation in the riparian zone and the surrounding forest decomposed in the reservoir and affected water quality. From the beginning, the municipalities challenged the BCER's dam on the grounds of possible pollution and renewed calls for changes in water treatment in 1907 and later in 1913, following a series of studies on water chemistry.[56] It is not unimaginable that the fishways – one potential source of pollution – were made inoperative to allay a controversy over water purity. At the very least, the fishway, once found to be ineffective, was not improved. One form of dam criticism may have been employed to quash another.

The BCER, however, did not buckle under the water-quality criticism. Its manager, Johannes Buntzen, simply bided his time until an opening appeared to turn the criticism to advantage. In 1908, after public complaints about water purity and dam safety as a result of a rise in the water level, the BCER used the criticism as an argument in favor of expansion. A bigger dam, Buntzen stated innocently to the press, would surely be safer.[57] Following an arduous political process, involving both the provincial and federal governments, BCER officials extracted from the federal government with jurisdiction in the railway belt the necessary licenses to proceed.[58] The license allowed for a new dam, an expanded pipeline, and a larger powerhouse, on the condition that the foundation was secured and

55. PSCA, File 1180.1–12, "Coquitlam Dam," W. D. Burdis to C. B. Sword, March 19, 1906.
56. UBC, BCER Papers, Box 74, File 1474, "Coquitlam Dam."
57. CSB, "Coquitlam and the Dam," *Columbian*, May 18, 1908.
58. Roy, "The Fine Arts of Lobbying," pp. 241–3.

that an engineer could monitor water quality.[59] In a time-honored tradition of backroom lobbying and taking advantage of jurisdictional rivalry, the BCER utterly outflanked the local municipalities.

In the process, the fishway, neglected and forgotten, was swept aside in the expansion. The new dam planned for Coquitlam Lake was to be substantially higher than the original, standing 70 ft at its tallest point from the riverbed. If the fishway in the first dam was unsuccessful, how was this new dam to be modified to allow fish passage? The problem seemed insoluble to the federal Department of Marine and Fisheries. R. N. Venning, the Dominion superintendent of fisheries, wrote to C. B. Sword, the BC inspector of fisheries, in the fall of 1909 to ask whether the fishway question should be put to rest. He had received representations, Venning wrote, that the fishway did not work and that whatever fish had passed needed to be removed after spawning because of the risk to the purity of the municipal water supply. Given the height of the dam and these other calls on "the general interest," he asked whether any further action was prudent.[60] Sword, the fisheries man on the spot, believed not. Although he thought that the past performance of the fishway was better than generally supposed, he did not think that its future held much promise or that the dam should be held up to accommodate fisheries concerns. In the grand scheme of things, the run was minor, whereas the dam was important. In any event, he wrote, coho salmon "are apparently not very particular where they spawn." Perhaps they would simply migrate elsewhere.[61] The views of Sword were probably not out of line with local canners. No record of protest exists surrounding the expansion and removal of Coquitlam fishway. And John Pease Babcock, so conspicuous a presence in the Quesnel Lake controversy, made no entrance at Coquitlam Lake. Either the threat was too small or the power politics too overdetermined to attract these other players.

After the completion of the new dam in 1912, it was only a matter of time before the runs were depleted. In the fall of 1913, when

59. CSB, "BC Electric Power Scheme," *Vancouver News-Advertiser*, February 12, 1909; "Dam at Coquitlam is Sanctioned," *Province*, April 23, 1909.
60. PSCA, File 1180.1–12, "Coquitlam Dam," Venning to Sword, September 3, 1909.
61. PSCA, File 1180.1–12, "Coquitlam Dam," Sword to Venning, October 7, 1909.

the cycle of sockeye that had left Coquitlam Lake in 1909 returned, members of the Coquitlam band appealed for a special fishery, fearing that this one would be their last. Gathering signatures from local commercial fisheries operators in Vancouver, the natives petitioned the federal Department of Marine and Fisheries for fishing access and the right to sell their catch commercially.[62] As this was expected to be the last run, the department permitted the request.[63] In 1913, around the time of the discovery of the Hells Gate slides, members of the Coquitlam band stood below the Coquitlam dam, gaffing the last salmon of a soon-to-be-extinguished local stock. Remembering these two episodes in 1919, Parnell Keary, a resident of New Westminster, wrote a letter to the editor of the *Vancouver Sun*, stating that native fishing had not diminished the Fraser's salmon: The white man's barricades were to blame.[64] It was a clever inversion of the usual rhetoric. Yet the BCER, the builder of the Coquitlam barricade, had no concern for such carping: Municipal critics were in their place, the fisheries department was entirely cooperative, and neither the cannery interest nor local natives posed a threat. This was now a power lake, materially and politically.

No similar disputes marked the development of other lower-basin rivers. In part, this was due to the location of the next phase of development on the Stave Lake system. Stave Lake sat above an 80-ft-high set of falls, well beyond the reach of any migratory salmon species even before a dam was built.[65] From the perspective of the commercial fishery and the fisheries regulators, the project would not harm salmon spawning habitat. The site was also further from the urban region than the Coquitlam River and remained beyond municipal water claims. Although the dam flooded traditional gathering sites of native groups, particularly the Kwantlen band, and affected their fishing success on the lower river because of a changed flow regime, there is no evidence that the Stave Lake Company concerned itself with these problems.[66]

---

62. CSB, "Permission Sought by Indians to Fish," *Vancouver Sun*, April 23, 1913.
63. CSB, "Indians May Catch Salmon," *Columbian*, May 8, 1913.
64. CSB, "Salmon Horde Perish In Effort to Pass White Man's Barricades," *Vancouver Sun*, October 12, 1919.
65. PSCA, File 1180 1–15 "Obstructions – History – Dams – Lower Fraser," April 18, 1940, p. 4.
66. Carlson, *Historical Atlas*, Plate 41, pp. 122–3.

Damage to salmon habitat did result from the later development of the Stave in the 1920s when projects linked watersheds
to increase flows for power generation. When Alouette Lake was
dammed and diverted into Stave Lake, for example, it cut off
salmon habitat and disturbed spawners in the lower Alouette River.
At the time of construction in 1923, fisheries officials established
that the Alouette watershed was used by five species of Pacific
salmon.[67] However, the BCER did not install fishways, and the federal Department of Marine and Fisheries and canners organizations
raised little protest. On the assumption that the dam would flood
spawning grounds, Chief Inspector of Fisheries J. A. Motherwell
decided against insisting on fishways to the BCER, judging the situation hopeless. "Under the circumstances," Motherwell wrote to
Babcock in 1923, "it has been decided that no further action should
be taken by this Department and that there should be no obstacle
placed in the way of the proposed development."[68] In the subsequent decade, the salmon runs on the Alouette fell off markedly;
attempts to relocate thousands of spawners in 1927, the year of the
dam's completion, did not succeed.[69] Over the next half-century,
spawners in the lower river survived in diminished numbers, facing
the erratic schedule of water releases based on waterpower principles, not natural runoff. The Alouette case suggested that, by the
1920s, the fisheries question had not gained increased importance;
rather, it had decreased as a factor in river development. There is
little evidence to suggest that the threats posed to salmon fisheries
figured significantly in planning decisions about dam construction
in the lower basin or beyond.

In retrospect, the lack of sustained concern for the effects of
development on the salmon fisheries may appear negligent or unusual, but the Fraser River case was not unique. To the south,
in the Columbia Basin, dam development proceeded on numerous tributaries in the 1920s and the main stem in the 1930s with
little knowledge of consequences for salmon or with little effective protest within government agencies or the fishing industry.

67. PSCA, File 1180 15, "Obstructions – History – Dams – Lower Fraser," p. 5.
68. BCARS, GR 435, BC Department of Fisheries, Box 63, File 590, Motherwell to Babcock,
    November 5, 1923.
69. PSCA, File 1180 15, "Obstructions – History – Dams – Lower Fraser," p. 5.

American historians have pointed to the significant barriers to cohesive defence on the Columbia born of racial animosities and divisions within the fishery based on differences in gear types and perceived effects on catch levels.[70] The Fraser case revealed some of these same divisions, but it also suggests the significant influence of private capital in the utilities field and the support for electrification as a general public good. One provincial official summed up this view in the late 1920s in the course of assessing the possibility of a dam on a large salmon river on the coast. Power was consumed generally, he observed; fish were exported and eaten by a minority. The government would have to seek the greatest good for the greatest number. That meant power, not fish.[71]

Water and the State

Although the federal and provincial governments paid little attention to protecting salmon spawning habitat, they did make attempts to harness the promise of water development. Although at the turn of the century no formal surveys of BC waterpowers or rivers had been undertaken, by the 1910s the federal government had initiated river surveys under the Department of the Interior and the Commission of Conservation. The aim was both to provide baseline information about flow regimes so that water conflicts in the federally controlled railway belt in BC might be mediated and to create a nationwide inventory of water resources in the manner of the U.S. Geological Survey.[72] In the 1920s, the provincial Ministry of Lands and Forests created a water-survey branch to conduct inspections of promising water-development opportunities. In the late 1920s and early 1930s, the provincial survey program catalogued a host of development sites in the middle and upper Fraser Basin.[73] Water surveys were published in handsome, promotional

70. Taylor, *Making Salmon;* White, *The Organic Machine.*
71. BCARS, Department of Lands and Forests, 'O' Series, File 001689, MacDonald to Howe, November 27, 1929. The river under consideration was the Nimpkish on Vancouver Island.
72. On the origins and early history of American state hydrology, see Follansbee, *History of the Water Resources;* Shallat, *Structures in the Stream.* For the international history, see Biswas, *History of Hydrology;* White, *Water Powers,* p. 1.
73. Farrow, "The Search for Power."

volumes designed to catch the eye of international investors with photographs emphasizing rushing rivers and graphs illustrating massive untapped potential. The threat that water development posed to the provincial fisheries went largely unmentioned.

This early phase of river survey helped to imprint a new image of the river. It drew on international developments in hydrological survey techniques, mimicked programs in the United States, and spoke an international discourse about water in cubic feet per second. C. R. Adams, an engineer with the U.S. Geological Survey, led some of the first federal surveys in BC.[74] Across the river basin, gauges were established to provide a serial record of river flows. In the 1930s, snow surveys began, based on a technique developed in Nevada, in order to improve river flow forecasts.[75] The surveyors were a diverse group with connections around the world. They brought experience gained in railway surveys in the Canadian North and Prairies, as hydroelectric engineers in California, and as hydraulic engineers in the service of the British Empire.[76] It was not out of place in BC water surveys to find local rivers compared to the Congo.[77] BC was linked, in short, to broader trends around the globe in the age of hydraulic imperialism that recast the meaning and purpose of rivers and imported technical expertise to convert rushing water into an industrial prime mover.

But however much this survey program paralleled and sought to mimic developments elsewhere, it fell short of state-led development programs, pioneered in the British Empire and Western United States.[78] Under British rule in Egypt and India, for example, dam and canal development had proceeded since the late nineteenth century under imperial auspices as the proper projects for government. In other British settler societies, like Australia, the imperial examples were crossed with American entrepreneurial influence to produce a series of large, state-run projects before the

74. White, *Water Powers*, p. 306.
75. Farrow, "Snow Surveys for Forecasting Streamflow"; idem, "Snow Surveys: A New Medium"; idem, "Forecasting Run–Off"; Mergen, "Seeking Snow."
76. Andrews, "Major Richard Charles Farrow"; BCARS, Ad MSS 392, Frank Swannell Papers, biographical information; Sparks, "British Columbia Water Surveys."
77. BCARS, Ad MSS 392, Frank Swannell Papers, surveyor general to Swannell, May 13, 1920, contained in 1920 diary; Meurling, "Description of Work," p. 145.
78. Headrick, *The Tentacles of Progress*, pp. 171–208; Tyrrell, *True Gardens of the Gods*; Worster, *Rivers of Empire*.

1930s. To the south, British Columbians observed the interventions of the U.S. federal government in western American water projects through both the Bureau of Reclamation and the U.S. Army Corps of Engineers. With the dawn of the New Deal in the 1930s, the role of the state only increased in large American water projects. The Tennessee Valley Authority (TVA), the Columbia River dams, and the massive Boulder (later Hoover) dam on the Colorado provided examples of the power of the state to confront and transform large rivers and to display its majesty in concrete edifices. Although Canadian newspapers were filled with envious descriptions of these American projects and noted others abroad, neither the provincial nor the federal government entered the water-development field in BC.

The limits of state involvement in water projects in BC were related first to the reluctance of governments to intervene in the economy and second to the biases of Canadian federalism.[79] At the provincial and federal level, the state was in a period of retrenchment. Across Canada in the 1930s municipal and provincial governments faced near bankruptcy and the federal government's debt ballooned. Although some attempts were made to provide public work projects and relief camps, the effects were limited. Nothing like the countercyclical programs of the Roosevelt administration in the United States occurred in Canada. The U.S.'s New Deal, to paraphrase John Herd Thompson and Stephen Randall, contrasted Canada's No Deal.[80] In the early 1930s, BC's provincial water surveys were cut along with other programs. Faced with a depression, the provincial government sought to reduce costs, not spend. Even a purportedly interventionist Liberal government in the late 1930s could not imagine massive deficit spending. Added to this, the province felt strongly about the proper bounds of its authority and resisted federal attempts to override its jurisdiction over resources. At the federal level, similar concerns about the effects of intervention in the economy limited state programs. In 1935 the Bennett Conservatives released some federal funds for water development in the Prairie Provinces, but only as part of a general agricultural rehabilitation package. In BC, with no pressing

79. On the political history of this period, see Thompson with Seager, *Canada 1922–1939*.
80. Thompson and Randall, *Canada and the United States*, p. 134.

drought to command national political interest, the prospects of water development for future growth weighed not at all in federal analyses of fruitful fields for investment. The Prairie case, in any event, had complex jurisdictional aspects. The federal government had no intention of mimicking the American example of a strong federal water-development agenda and of entering more federal–provincial battles over resource control. Canadians could watch American projects rise with envy, but within the limits of their state system and the political culture of economic responses to the depression, they could do little more.

In part the differences of the state role in water development in BC can also be traced to the limited range of intended resource uses at this date. State support in BC sought to promote water for power; but the prospects of irrigation or the need for flood control were rarely mentioned. The Fraser Basin contained limited arable land, located mainly in the delta and valley near the coast and along narrow strips of land on terraces in the middle and upper basin. Other parts of the basin contained valuable grazing territory, but on the whole the basin was too rocky, its soils too inhospitable and its steep hillsides too covered with coniferous forests to foster any serious notions that the Fraser Basin could become Canada's version of California's Imperial Valley. Further, although periodic floods occurred on the Fraser, there had been no serious inundation since 1894. With provincial funds limited during the 1930s, there were no designs to improve the river or expand existing dikes in the Fraser Valley. It would not be until 1948 that a major flood inspired the provincial and federal governments to consider new means to tame the river. But before that time, the limited state role in the water-development field targeted one promotional focus: hydroelectric-power development. This placed BC in a different comparative position to the western United States where dam development contained a variety of goals for transforming rivers and was backed by diverse constituencies. In the Columbia Basin, for example, the two key federal agencies involved in main-stem dam projects, the Bureau of Reclamation and the U.S. Army Corps of Engineers, looked to problems of irrigation, navigation, and flood control. Hydroelectricity was viewed as an important and integrated adjunct to these other aims, but was not the only focus.

Conclusion

In 1929, a group of fisheries scientists and officials from Canada
and the United States met under the auspices of the International
Salmon Investigation Federation, a new organization aiming to create fisheries research links on the Pacific coast. They discussed the
relationship between dams and fish. In the U.S. Pacific Northwest,
dam projects on the Columbia were about to commence. Canadians wondered whether or not similar development plans might be
under way in BC. J. A. Motherwell, the chief inspector of fisheries in
BC, warned that it was entirely possible that power interests would
seek to dam Hells Gate at its narrow gorge and to transform it into
the keystone of power development on the Fraser River.[81]

Before this time, damming the Fraser River had remained an
unrealized fantasy, out of touch with the realities of local market
demand, the costs and technologies of transmission and the corporate strategies that favored cautious investment and concentration
of development. When hydroelectricity arrived in BC at the turn of
the century, developers had sought rivers and lakes close to the major urban market in Vancouver, cognizant of the limited potential
of demand in the short term and the expense and impracticality of
damming the Fraser's main stem. By managing potential rivals and
the scope of state regulation and competition, the BCER had carved
out a stable investment climate in which it could meet rises in demand incrementally by intensifying development at existing sites
without indulging in extensive development or risky long-distance
transmission. The business history of early hydropower development implanted the tributaries of the lower basin with dams, but
left the main stem for future investigation.

Dam developers, governments, and promoters had demonstrated
little concern or attention to the effects of dams on salmon in this
early development phase. At Quesnel Lake, a speculative mining
dam had cut off one of the more important sections of salmon
spawning habitat in the basin. Affected interests had to spur governments to engage their legal powers to regulate dams and modify them to protect salmon habitat. But even this limited form of

81. BCER CF, "Our Salmon Streams," *Province*, April 8, 1929.

regulation and intervention could be compromised. In the lower basin, fisheries officials effectively surrendered salmon streams with little protest, in view of the public interest in industrial and power expansion in the Vancouver region. Fisheries interests protested some of these shortcomings but failed to engage in unified action or explain clearly why the protection of salmon habitat should trump the power ambitions of Canada's Pacific metropolis.

Power and dam development proceeded in BC with state approval but little financial support or direct intervention. Whereas in parts of the British Empire and the western United States, the state had added water-development projects to the list of state responsibilities in capitalist societies, neither the Canadian federal nor the BC provincial state chose to build projects or support established ones. Water development remained within the private sphere. Viewed comparatively, the electrification of BC remained behind that of neighboring areas of the United States. The BCER's corporate strategy contrasted state intervention markedly by following demand rather than seeking to catalyze it.

In 1919 federal officials believed that the Fraser would not be dammed primarily because of the threat to salmon, but the history of early dam development in the Fraser Basin would suggest otherwise. Danger to the fishery did not stop or significantly delay a single development project in the Fraser Basin before the 1940s. The Fraser remained without dams on its main stem because of the limits of the river for lower-basin development, market and business conditions, and the lack of state involvement – not because salmon held priority in government or society.

3

Remaking Hells Gate

The rough water is at Hells Gate. More than two decades after the Hells Gate slides, in the summer of 1938, Bill Ricker, a scientist with the International Pacific Salmon Fisheries Commission (IPSFC), perches on the rocks and investigates the causes of the precipitous decline of Fraser sockeye. The remnants of the slides lie on the opposing bank (see Photograph 6).

A copy of the photograph is among the papers of William Thompson, who, in 1938, had recently assumed the directorship of the Salmon Commission's scientific investigations after a distinguished career with the North Pacific Halibut Commission and as chair of the University of Washington's College of Fisheries. Unlike Ricker, who left after his first year of study, Thompson devoted the better part of a decade to Hells Gate; his ideas about its role in obstructing salmon migrations would provide the rationale for the construction of fishways at this point in the mid-1940s as one prong of a major effort to restore the salmon runs. After the completion of the fishways, when Thompson set down his ideas about Hells Gate for scientific scrutiny, his early charge, Bill Ricker, would criticize them strongly, engaging in a prolonged controversy with Thompson that would come to involve the reputations of their respective scientific institutions and national fisheries science communities. But in the summer of 1938, none of these later controversies could be imagined. Ricker leaned over the edge, photographing salmon, and the lens captured him too.

The IPSFC's research program originated in the politics of the Pacific salmon fishery. Since the turn of the century, Canada and the United States had debated the division of the catch and engaged

Photograph 6. Dr. William Ricker at Hells Gate, 1938.
(*Source:* "Dr. William Ricker at Hells Gate, August 19, 1938." Photo taken by A. J. Tubb. University of Washington Archives, William F. Thompson Papers, Acc. 2597-3-83-21, Box 9, File Photos.)

in periodic negotiations to pursue international fisheries management. Agreement was obstructed, however, by three major factors: constitutional disputes in the United States over states' rights and the fishery, conflicts within the fishing industry over catch levels and gear restrictions, and unequal bargaining positions created by the U.S. majority share of the fishery. Not until the 1930s, after sharp declines in the fishery, the end of the American trap fishery, and a shift in salmon migration patterns that favored Canadian fishers, did the United States and Canada successfully conclude a joint management and catch agreement – the Pacific Salmon Convention of 1937. The convention created the IPSFC to carry out its mandate and conduct an eight-year scientific survey of the resource with a view to future regulation.[1] In the course of general surveys of the Fraser sockeye in 1938, commission scientists discovered blockage conditions at Hells Gate.

Political, social, and natural factors conditioned scientific knowledge of Hells Gate. Scientists' views were developed in the context of the national and institutional politics of the Pacific fishery and the IPSFC; their research practices involved a complex cultural and natural selection through interaction with native fishers and tagging methods, and their justification and criticism of the fishways subjected research to personal, institutional, and national divisions. In this chapter, I ask how scientists remade Hells Gate, how they tried to understand this place, and how they debated its meanings.[2]

Establishing a Research Program

Reflecting on the early years of the IPSFC, William Thompson noted in the late 1950s that the eight-year period when the commission was devoted solely to scientific inquiry allowed for uncommon latitude in charting new directions in fisheries research. "The Commissioners were, for a time, free from the job and glory seekers who were not interested in doubtful personal futures . . . free from the demands of regulation according to this or that popular theory [and] . . . free from the pressure of immediate results." The treaty, he argued, provided a research opportunity beyond the clawing control of "small organizations" and national policy concerns.[3]

1. On the background and history of the 1937 Convention, see Dorsey, *The Dawn of Conservation Diplomacy*, Chapter 3, pp. 76–104; Taylor, "Historical Roots"; Tomasevich, *International Agreements*.
2. A number of previous studies have examined the remaking of Hells Gate in the 1940s: Fisheries scientist and historian John Roos provides a thoughtful record of the Hells Gate investigations in his history of the salmon commission, and fisheries scientist Tim Smith treats the Thompson–Ricker controversy as an opportunity to examine shifting ideas in the field of scaling fisheries: Roos, *Restoring Fraser River Salmon*; Smith, *Scaling Fisheries*.
3. University of Washington Archives (hereafter UWA), Acc. 2597-3-83-21, Thompson, W. F. Papers, Box 3, "Fishery Treaties Between the U.S. and Canada," nd (but probably ca. 1959), pp. 18–22.

Memory may be a salve, but Thompson's remarks do provide insight into the importance of institutional arrangements for the conduct of scientific research, particularly when such inquiry intersects with vested economic and political interests. Whether the IPSFC was as successful at deflecting industry pressures and national policy concerns as Thompson remembered is another question.

The commission established under the Pacific Salmon Convention was composed of three layers of organization and operation. The commission proper contained six members (three from each country) and held responsibility for the general planning and implementation of the convention. Members of the commission were appointed by their respective national governments and were connected in some respect to the fishing industry or to regulatory bodies. The founding Canadian commissioners, for example, comprised a fisheries official (W. A. Found), an industry representative (A. L. Hager), and a politician (Tom Reid); two of the American commissioners, on the other hand, were fisheries officials (B. M. Brennan and Charles E. Jackson), and the last was a prominent Washington lawyer with expertise in fisheries matters (E. W. Allen). An advisory committee, made up initially of ten industry representatives, five from each country, performed an ad hoc role connecting commissioners to industry and organizational concerns. A third layer consisted of professional and technical staff overseen by a director of investigations, separate from, but subordinate to, the chairman of the commission.[4] This group, the most important in identifying and carrying out the restoration efforts in the commission's early years, is the focus of my analysis.

The scientific activities of the early commission derived their main impetus and direction from William Thompson. Thompson might be considered as one of the leading lights of the second generation of fisheries scientists on the Pacific coast in the twentieth century. A product of the Stanford fisheries program, he undertook important fisheries studies as the head of the North Pacific Halibut Commission in the late 1920s and developed a substantial research career modeling fisheries. He also acted as chair of the University of Washington College of Fisheries starting in 1930 and oversaw

4. On the organization of the commission, see Roos, *Restoring Fraser River Salmon*, pp. 54–5.

the transformation of the school from a practically oriented program to an important research institute that attracted significant
private, state, and federal research funds.[5] A better scientist than
politician, Thompson probably accepted the role as director of investigations under the Salmon Convention more for the research
opportunity than for the prestige. He did not enjoy publicity and
resigned within five years, embittered by the personal and political
conflicts that had tainted his scientific mission. During those five
years he had led a research team to one of the most important single
discoveries in fisheries management in BC history.

Thompson's responsibility was to gather and supervise a team of
researchers to pursue a set of investigations with a view to restoration and regulation of sockeye salmon runs. Building an international scientific institution, however, also involved problems of national and international politics. To the consternation of Canadian
commissioners, Thompson initially tried to base the work at the
University of Washington (UW). Canadian MP and commissioner
Tom Reid insisted that the benefits of the convention ought to be
more evident in Canada, particularly his riding of New Westminster. He also criticized Thompson's penchant for hiring Americans,
and he asked why Canadians were not being hired for the research
jobs. American commissioner B. M. Brennan replied that there simply were no qualified University of British Columbia (UBC) grad-
uates; better people, trained in fisheries science, were available in
Washington.[6] The disagreement signaled Canadian fears that the
commission would simply become an American research effort;
from the opposing perspective, American commissioners believed
that Reid was bullying Thompson and "look[ed] upon the commission as an opportunity for patronage."[7] Thompson lost the
battle over the location of the commission offices, but managed to
carry out substantial research in the UW labs and to hire his own

---

5. UWA, Acc. 1683-71-10, Van Cleve, Richard Papers, Box 4, R.V.C., "The College of
   Fisheries, University of Washington," nd, pp. 1–2.
6. UWA, Acc. 2597-3-83-21, Thompson, W. F. Papers, Box 7, File 1940, B. M. Brennan to
   A. J. Whitmore, May 31, 1940 (copy). Despite Thompson's preference for University of
   Washington-affiliated staff, he did, Brennan reported, advise UBC on how to improve its
   undergraduate program to bring it into line with the commission's requirements.
7. UWA, Acc. 129-3, Allen, E. W. Papers, Box 2, File 2–52, E. W. Allen to Charles E.
   Jackson, U.S. Bureau of Fisheries, July 1, 1939 (copy).

people. Some of these turned out to be distinguished Canadian researchers. Jack L. Kask, a UBC graduate and UW PhD, who had formerly worked under Thompson on the halibut study, as well as Bill Ricker and Russell Foerster, two Canadian researchers based at the Nanaimo federal fisheries station, joined the IPSFC in its first year. Ricker would conclude his relationship with the IPSFC a year later and, in the late 1940s, become its most outspoken scientific critic.

Rarely are such large-scale subjects like the sockeye salmon of the Fraser Basin examined in integrated studies. Before the advent of the IPSFC research program, a range of provincial and federal fisheries scientists studied Fraser River salmon. Much of the early work had cataloged distributions and had sought to determine the validity of the home stream theory. Stanford zoologist Charles Gilbert had produced the most important work in this line by attempting to distinguish racial groups within species by means of growth-ring analysis. In the 1920s, federal research made advances in the study of fish culture with a set of intensive studies at Cultus Lake on the returns of "wild" and reared sockeye. Other work concerned the control of predator populations. All of these projects were basic in different ways, but they were regionally segmented and of limited application. The IPSFC research mandate allowed Thompson and his team to look to broader questions that connected the watershed as a whole and to suggest the basis for a sweeping restoration program. Other than the Halibut Commission and the pioneering International Council for the Exploration of the Sea, a body that coordinated fisheries research in the North Atlantic, few precedents for such a program existed internationally.[8]

Thompson's research program aimed to maintain flexibility.[9] Research funds were not allocated to particular projects of 5-year duration; instead, IPSFC scientists pursued a problem-oriented survey approach in the first year to identify worthy topics of study. Although the existing literature on the Fraser fisheries was collected

8. For a discussion of the Halibut Commission, see Tomasevich, *International Agreements*, pp. 125–209; on the science of the International Council for the Exploration of the Seas and the Halibut Commission, see Smith, *Scaling Fisheries*, pp. 110–229.
9. Thompson's program is detailed in a memorandum to the commissioners: UWA, Acc. 2597-3-83-21, Thompson, W. F. Papers, Box 7, File 1938, Thompson to the IPSFC, May 18, 1938.

and supplemented with historical material on past catch levels, the main emphasis of the initial field work was to differentiate the sockeye fishery by tagging studies and spawning bed surveys. Early studies of Gilbert and others at the turn of the century had established that salmon populations were not homogeneous, but could be separated into races with their own migration patterns and spawning areas. The racial theory was a cornerstone of the IPSFC's initial survey. By tagging sockeye in closed periods off the coast of Vancouver Island near Sooke and at various stages upstream, scientists collected statistics on the length of migration and characteristics of the particular racial groups. Survey parties also examined each of the major spawning areas in the watershed during the summer and fall runs, determined the number of returning spawners, and entered relevant environmental information into standardized notebooks for each spawning region. Incidental to this work, some attempt was made to observe the native fishery in order to develop some estimate of its annual take with a view to regulation. And despite some initial reservations, Thompson decided to fund the continuation of the Canadian Department of Fisheries studies at Cultus Lake on the rate of sockeye returns.[10] This marked the transition from federal to commission control in Fraser sockeye research. Hereafter the field would be dominated by the commission while the federal department would turn its attention to the Skeena River.[11]

In describing his approach, Thompson wrote, "I am holding the program open to change. It must not be allowed to crystallize before the direct utility of its several features is seen."[12] After the first summer of investigations, some of those "features" were becoming apparent. One was the obstructions in the Fraser Canyon. In some sense, the problem was stumbled on. Tagging experiments in saltwater had determined that too few of these marked fish were surviving the entire migration process to provide meaningful statistical data. Various upstream locations were chosen for tagging fish and collecting in-river migration data. One such location near

10. The IPSFC's research program is outlined in Thompson's memorandum to the commissioners: Ibid.
11. Johnstone, *The Aquatic Explorers*, pp. 175–6.
12. UWA, Acc. 2597-3-83-21, Thompson, W. F. Papers, Box 7, File 1938, Thompson to IPSFC, May 18, 1938.

Yale was shifted to Hells Gate in midseason because fish were more easily captured there. From rocks and crags, and later from small scows, scientists fished with gill nets for sockeye, tagged them, removed some of their scales for racial analysis, and then released them. Findings based on this method would shape the course of the IPSFC's research mandate over the next decade.

Tagged fish, the scientists found, did not pass Hells Gate as expected. Frequently they were held up for days, turning up in the tagging nets more than once as recaptures, and sometimes downstream, as far away as the river's mouth. Although some of the tagged fish were recaptured upstream and did provide evidence of timing of migration to spawning areas, enough did not get through that Thompson and his team thought it prudent to focus more attention on the problem in subsequent seasons. Could it be, they asked, that the rumor about Hells Gate, so frequently dismissed by fisheries officials over the past two decades, was true? Did the gorge still contain material from the slides that made salmon passage difficult? From 1939 to 1941, the IPSFC placed a special emphasis on answering this question. Their main means of analysis was the tagging procedure, contextualized by relevant data on water levels, catch statistics, and spawning ground counts of escapements. But, given the centrality of the tagging method, it is well to consider the operation of this experiment more closely. How were small celluloid disks representative of shifts in nature?

Fishing for Tags

By the 1920s fish-tagging experiments were becoming a fundamental tool in large-scale fisheries studies. Thompson had used them in the Halibut Commission work, and they had been widely used in large oceanic studies.[13] Joint Canada–U.S. tagging experiments on salmon had provided part of the conceptual basis for the Salmon Convention by demonstrating the transnational migration patterns of Fraser sockeye. But before the commission experiments

13. On early plaice-tagging experiments, see Smith, *Scaling Fisheries*, pp. 143–6. For an overview of marking studies in fisheries biology, see McFarlane, Wydoski, and Prince, "External Tags and Marks."

Photograph 7. Tagged sockeye salmon from the Hells Gate investigations. Notice the placement of the tag below the dorsal fin.
(*Source:* Pacific Salmon Commission Collection.)

on the Fraser, tagging had been used most frequently to demonstrate the migratory paths of fish in the ocean. It had never previously been used in a major study on the Fraser. Wilber Clemens of the Biological Board station in Nanaimo had suggested such a project a decade before during a Board of Engineers investigation of Hells Gate, but without result.[14] In-river tagging was adopted by the commission only when it appeared that too many of the fish tagged in saltwater were being taken by the commercial fishery.

Fish tagging was a scientific exercise in differentiating populations and analyzing their movement through space. In the commission experiments, fish were captured according to a random fishing process, pierced with a nickel pin, and identified by two celluloid disks, inscribed with a serial code, placed directly under the dorsal fin (see Photograph 7). The fish were then returned to their natural habitat, and scientists waited to discover where they reappeared. The assumption was that the tagged fish mirrored the experience of the larger population, at least in probability terms. Tags did not intrude on or alter natural patterns; they reflected them.

14. NAC, RG 23, Box 679, File 713-2-2[9], J. A. Motherwell to W. A. Found, April 19, 1928.

Tagging, however, was not carried out in a hermetically sealed scientific space in which natural relationships could be distinguished unproblematically from cultural contexts or ways of seeing. Data were meant to provide direct clues about natural change and salmon movement, but the very means of collecting tags placed filters between the scientist and the rest of nature. Collection methods introduced various forms of selectivity. The very tools of capture were selective: Gill nets, which snagged certain sizes of fish more than others, were replaced with dip nets in 1942.[15] More fundamentally, such nets were imprecise gauges of passing populations. The disjuncture between an ideally constant tagging pressure and a variable rate of salmon passage meant that, when a large cohort passed or was delayed, a different proportion of the population was sampled than at other times.[16] What this meant for the nature of the sample and the resulting data was unknown. Beyond the gate, in the upper-basin spawning grounds, the collection of the data created further problems. All tags were not retrieved. On some streams, river flow carried the carcasses of spawned fish away, taking their precious tags with them.[17] On others, tags were discovered, but only after the spawning was complete. Judging when the fish arrived became a guessing game that was only compounded when intermediaries turned in the tags. There were oddities that could not be wholly explained: Many of the tagged fish collected in the spawning grounds bore scars. Thompson stated in his final report that the proportion of fish thus affected was "relatively high." He suggested that native gill nets might be the cause.[18] But, as with all of these anomalies, it was difficult to say. Commission scientists knew these problems existed and acknowledged them in their published findings. They did not, however, publish quantitative analyses of how much of the data might be affected by any or all of these discrepancies. Such problems would provide the basis for subsequent critiques of the commission's science and of its conclusions.

Perhaps the most blatant problems with the data collection showed up in the interaction of commission scientists and native fishers. On the face of it, the commission's plan to study native

---

15. Thompson, *Effect of the Obstruction*, p. 97.   16. Talbot, *A Biological Study*, p. 12.
17. Talbot, *A Biological Study*, p. 22.   18. Thompson, *Effect of the Obstruction*, p. 98.

fishing on the Fraser blended perfectly with its tagging experiments. Natives were asked to return fish tags taken in the seasonal fishery to the commission. This would provide scientists with data about fish movement as well as about fishing pressure. As with commercial fishers in saltwater experiments, native fishers would be paid $.50 per tag. A simple arrangement no doubt, but one complicated by the long history of antagonism and unequal power relationships between native fishers and fisheries regulators in the canyon. Native peoples did not return the tags as expected, but sometimes saved them or turned them in far from the point of catch. The problem was not that native fishers were necessarily setting out to sabotage research, but that they were collecting tags for their own reasons. One person's data were another's $.50.

Whereas the tags were a marker of fish passage for scientists, inscribed with data and representative of natural change, they became a "fungible" in the economy of native fishers in the canyon. Karl Polanyi defines a fungible as a durable object that can perform the functions of money – as a means of payment, a standard of value, a store of wealth, and a means of exchange.[19] Although no statistics were published by the commission about the number of tags collected by natives, in the 1941 season Thompson estimated that, because of a lack of commission tag collectors on the spawning grounds, over $1000 would be paid to natives searching the spawning grounds alone.[20] That works out to 2000 tags, more than 10% of the total number of tags used in that year. In the seasonal fishery, tag collection became a lucrative sideline, and sometimes an end in itself. Fish tags turned into local currency.

The fungible quality of fish tags was a lesson that commission scientist Jack L. Kask learned with much frustration in the fall of 1940.[21] After recording peculiar patterns of tag returns in the canyon that did not correspond to commission expectations, Kask was sent to investigate how tags were collected and returned. At

19. See Karl Polanyi's essay on "Money Objects and Money Uses," in *The Livelihood of Man*, ed. Harry W. Pearson (New York: Academic, 1977), pp. 102–3.
20. UWA, Acc. 2597-3-83-21, Thompson, W. F. Papers, Box 7, File 1941, Thompson to IPSFC, August 4, 1941.
21. PSCA, File 2550.2-21, J. L. Kask, "Indian Fishing for Tags in the Closed Area Above Hell's Gate," November 4, 1940.

the Indian reserve near Anderson Creek, Kask questioned Chief Joe Brown about tags and discovered, to his displeasure, that tags were captured by a variety of people, some without fishing permits, who subsequently took them to places as far away as Lytton before returning them, if they did so at all. Besides confiscating some illegal gaffes that he found in the vicinity, Kask collected eight tags from Chief Brown and tried to insist on the importance of prompt tag returns. Kask was one of a long line of federal salmon officials who told native fishers of the canyon how to fish and expected their cooperation. He shared his frustration and prejudice in a memorandum to his superiors: "A thorough search of the Indian villages would probably unearth many more [tags], although the Indians do not hand in the tags until they are good and ready and as long as there are stores and other centres where cash can be obtained for tags it will be difficult for any commission employee to get to them."[22]

Stores accepting tags? This was a key problem, said Kask. The commission had hired a scattering of individuals in the canyon to collect tags directly from native peoples and record these returns promptly. Commission scientists did not envisage the emergence of middlemen. To study the methods of tag collection, Kask accompanied one such commission employee, Tom E. Scott, a retired federal fishery agent based in Hope, during his round in the lower canyon.[23] While insisting that he collected all tags directly from native fishers, Scott led Kask to several general stores. At Yale, he confessed that the majority of the tags from the lower canyon ended up in the cash register of the local Chinese–Canadian shopkeeper. Natives used tags in the store as cash equivalents. The proprietor held the tags, and Scott reimbursed him for the stated price of $.50 per tag. Or that is what Scott said. After visiting the Spuzzum general store where a similar transaction occurred and then Alexandra Lodge where dealings were carried out beyond Kask's view in the kitchen, the commission scientist had a fair idea of how the wily Scott operated. "Scott's great enthusiasm for collecting tags can be explained in this way. In 1938 and 1939 tags were redeemed at his

22. Ibid.
23. Idem, "Trip Made with Thomas E. Scott to Recover Tags and Remove Weirs in Nicola Valley," November 4, 1940.

appointed centres of tag collection at a reduced rate. As they were turned in to the Commission at the full rate of 50 cents per tag, it is conceivable that a small rake-off was made by the store-keeper and Scott."[24] In view of Scott's activities, said Kask, it would be best to stop employing tag collectors who used further middlemen and did not keep accurate records. A commission scientist, he argued, ought to be employed full time to ensure accuracy and prompt collection.

Because tags were stores of value for native peoples, they attracted a different kind of fishery: a strategically aimed fishery that bent "normal" fishing pressures in new directions and frustrated commission statisticians. A skilled fisher could see the shiny white disks with a red bullet at their center. Native peoples fished selectively for salmon because they contained use value in food and exchange value in celluloid.[25] Only in 1947 was a new kind of disk used that was less visible under water. In that year, commission scientists reported a significant drop in native tag catches.[26] Commission scientists also suspected that fish that were unable to pass Hells Gate and drifted downstream to die were monitored and collected by native fishers. Given the haphazard recording system for tags, at least in the first few years of tag collection, it is entirely possible that such tags were mixed up with those of different catch dates or carried north up the canyon and exchanged in a store beyond Hells Gate, giving commission scientists erroneous data from which to measure the passage of fish. Moreover, selective fishing may well have exaggerated the extent of the native fishery. For perhaps the first time since restrictions were imposed on native fishing after the Hells Gate slides, natives were reaping some material return from the regulatory process. In so doing, they were incidentally causing problems for the commission scientists.

Kask's advice affected future operations. Beginning in 1941, commission scientists collected tags directly from natives in return for the $.50 price. The middlemen were gone, and surveillance was

24. Ibid.
25. PSCA, File 2550.2-3, A. Welander and Peterson, "1941 Indian Fishery Report, Lower Fraser and Canyon."
26. Talbot, *A Biological Study*, p. 31.

intensified. In 1944 G. V. Howard wrote a guide for commission tag collectors that explained the best method[27]:

Visit all the Indian fishing stations in your district as often as possible, and acquaint yourself thoroughly with these localities. Acquaint yourself with these Indians and attempt to gain their confidence. In this way you will be able to determine the number of Indians who actually engage in fishing. From these fishermen obtain the following:

1. Name
2. Permanent Address
3. Number of dependents
4. Occupations other than fishing
5. Reliability of volunteered information

Collectors were instructed to record whether they collected such counts verbally or made them themselves, specify types of gear used, determine the placement of fish stations, and note how fish were preserved and consumed. Cards were kept on each fisher, and daily reports were filed. Salmon scientists *cum* ethnographers were attempting to gain a comprehensive sense of the native fishery, not only to control it and set limits on the catch but also to ensure the reliability of their data. Although native fishers experienced the most intensive surveillance of their fishery to date, commission scientists were coming to believe that their data were solid. Their ethnographic research was validating their tags as mirrors of nature. Of course, for native peoples the tags were still worth $.50.

Answers

After a number of field seasons, the commission scientists concluded that water levels were a primary cause of fish problems at the gate. The tagging experiments provided enough data on the time it took fish during periods of normal passage to pass the gate and turn up in spawning beds so that anomalies could be spotted. The major anomaly appeared in the recapture and upstream recovery

27. PSCA, File 2550.2-56, G. V. Howard, "Instructions for the Collection of Indian Fishery Statistics," 1944.

data. At water levels between 26 and 40 ft in Hells Gate, the number of fish recaptured below the gate after tagging climbed sharply, leading observers to conclude that few fish were passing. This seemed to be further substantiated by the low recovery of fish upstream after these "block" periods. The increases in recaptures showed a strong correlation with a middling water level in the annual fluctuation. They did not seem to correspond to other factors. The gate's unevenness underwater seemed to create high turbulence at certain levels and make passage increasingly difficult when the water dipped into the danger zone. It was as if the gate were shaped like an hourglass and fish were trying to pass – but failing – when water coursed through the narrow middle section.[28] By the beginning of the 1941 field season, commission scientists assumed that water levels were the primary problem, an assumption that an expanded research program in the upcoming season provided an opportunity to test.

The 1941 field season was as unusual as it was revealing. From early July until the end of October, Hells Gate appeared to be blocked to migrating salmon. In previous seasons blocks lasted for up to a week. In 1941, they lasted months. William Thompson, a scientist not fond of superlatives, was astounded and said so in his memoranda to the commissioners.[29] It was as if, he wrote in a later report, the whole drama of 1913 were being played out again in front of the scientists' eyes. Just as in 1913, when John Pease Babcock had surveyed the slide scene, salmon gathered in a confused traffic directly below the gate. They stretched down the river for 6 miles, and, as the season progressed, matured into the famous red of the spawning sockeye. Few passed through in the late summer months. Hardly any passed in September. For much of the season, water rumbled through the gate within the middling zone. A few respites in July, early September, and in late October allowed for some fish to pass through. Some of these fish were tagged, but few of them were discovered later on the spawning grounds.[30]

28. I borrow the metaphor of the hourglass from Northcote and Larkin, "The Fraser River," p. 196.
29. UWA, Acc. 2597-3-83-21, Thompson, W. F. Papers, Box 7, File 1941, Thompson to IPSFC, October 31, 1941.
30. Thompson, *Effect of the Obstruction*, pp. 92–96.

The spectacle of blocked salmon impressed the scientists and led to a rapid redeployment of scientific effort. At the beginning of the season Thompson had laid out a research program that included studies of the native fishery and of the long-term consequences of a dam built at Quesnel Lake as well as expanded work on tagging at Hells Gate.[31] But as the salmon migrants began to mount below the gate, all hands, including Thompson's, turned to Hells Gate.[32] By late August two teams of fish taggers handled 150 fish per day. Over 13,000 sockeye would carry tags by the end of the season. The project was, Thompson noted with pride, "one of the most extensive tagging programs of its kind ever undertaken."[33] Other projects risked incompletion, but the opportunity provided by unusual conditions had not been missed. Now the problem was to tie all of the data together: "Unmistakable as the indications are," Thompson stated, "the returns must be tabulated and analyzed with care."[34]

As the drama unfolded, Thompson thought he foresaw the process that lay ahead. The press was beginning to publish stories about the massive buildup and journalists wanted interviews.[35] By contrast, Thompson wanted, as he told the commissioners, to be "protected."[36] He did prepare a press release on the problems at the gate, but argued vigorously within the commission that the press coverage should not arouse alarm.[37] The risks were too great that publicity would force political decisions on the commission that could only disrupt the research. Conclusions, he advised Miller Freeman, the publisher of the *Pacific Fisherman*, were premature.[38] The commissioners acceded to his request.

31. UWA, Acc. 2597-8-83-21, Thompson, W. F. Papers, Box 7, File 1941, Thompson to IPSFC, August 4, 1941.
32. Ibid., Thompson to IPSFC, October 31, 1941.
33. UWA, Acc. 129,129–2, Allen, E. W. Papers, Box 3, File 5, Thompson to IPSFC, November 14, 1941.
34. UWA, Acc. 2597-3-83-21, Thompson, W. F. Papers, Box 7, File 1941, Thompson to IPSFC, October 31, 1941.
35. See, for example, Bruce Hutchison's epic article on the blockade: "International Sockeye Board," and the *Vancouver Sun* editorial and article of the same day, "Salmon Blockade," and "Salmon Board Declares Hells Gate Must Be Cleared."
36. UWA, Acc. 2597-3-83-21, Thompson, W. F. Papers, Box 7, File 1941, Thompson to IPSFC, October 31, 1941.
37. UWA, Acc. 129, 129–2, Allen, E. W. Papers, Box 3, File 5, Thompson to IPSFC, November 14, 1941.
38. UWA, Acc. 1038, Freeman, Miller Papers, Box 2, File 2–38, Thompson to Freeman, September 5, 1941.

Part of the reason why Thompson did not wish to attract attention to the apparent blockage at Hells Gate was that he thought the most likely solution required further study. To restore the gate and release the blockage conditions, a fishway of some kind would be needed. This was not a problem that could be handled quickly. The commission scientists were biologists, not engineers. New expertise would be needed in order to proceed. Furthermore, any building project would require a special disbursement from the national governments. That might not be simple to procure. Better, he thought, to control the flow of information as much as possible so the request for a fishway, when it came, would not be prejudged.

Fishways

Early in the century fishways were simple in design and crude in execution, but by the 1940s the technology had developed substantially.[39] These advances occurred principally as a by-product of the development of the Bonneville and Rock Island, dams on the Columbia River.[40] The Bonneville project included an extensive fishway and elevator system as an integral aspect of the design. Although it was unclear by the 1940s how well fishways operated over the long term, at least at Bonneville they appeared capable of passing fish. Thus the fishways at Hells Gate were imagined in an atmosphere where dams were the problem and fishways the technical solution.

In 1941 preliminary work at Hells Gate established that two "jutting rocks" on either side of the river created obstructions to fish and increased the fall of the water at the problem levels between 26 and 40 ft.[41] Creating safe passage would require the alteration or circumvention of these points. With a special one-time disbursement of $45,000 from the two national governments, the

---

39. For a brief review of the history of fishway designs, see Clay, *Design of Fishways*, pp. 14–18.
40. Pitzer, *Grand Coulee*, pp. 223–7; Smith, *Salmon Fishers*, p. 78; White, *The Organic Machine*, pp. 89–98.
41. This description of the engineering studies is based on Bell, "Report on the Engineering Investigation of Hell's Gate."

commission engaged a number of hydraulic engineers to study the problem and recommend a solution. These engineers drew on experience from the Columbia River dams and employed established river modeling methods following pioneering investigations by the U.S. Army Corps of Engineers during the 1930s.[42] At Hells Gate, Milo Bell, formerly of the Washington State Fisheries Department, took on primary responsibility for engineering investigations and contributed his experience gained as a designer of the Bonneville fishways.[43] At the UW, hydraulic engineering professor Charles W. Harris oversaw the construction and testing of a Hells Gate model with the assistance of Walter Hitner, also of the UW, as well as UBC engineering professor Edward Pretious.[44] At all points during their studies a team of biologists was at the ready to advise on the physiological and behavioral capacities of salmon.

By 1943 they had a prototype.[45] The fishways were unlike those previously created for dam structures that carried fish up and over obstructions. Instead they were designed to operate at different stages of the gate on both banks, assisting fish at only problem water levels. They would not surmount the gate, but rather would work through it. Positioned directly behind both of the "jutting rocks," the conduits would provide salmon with alternative routes around high-velocity points with a steep fall. At safer water levels, the fishways would be either submerged or above surface. Novel to the design was the use within the fishway flumes of vertical slot baffles within the fishway to slow the water speed to a consistent and manageable level.[46] There was a deliberate attempt to disturb the

---

42. Reuss, "The Art of Scientific Precision."
43. "Famed Engineer on Hell's Gate Project," *Vancouver Sun*, February 8, 1944.
44. UWA, Acc. 2597-3-83-21, Thompson, W. F. Papers, Box 7, File 1942, Thompson to IPSFC, May 27, 1942.
45. For a discussion of the hydraulic studies, see Bell, "Report on the Engineering Investigation."
46. Clay, in *Design of Fishways*, p. 13, describes the operation of the vertical slot fishway thus: "This fishway is constructed by installing a series of baffles at regular intervals between the walls of a flume. The baffles are so shaped as to partially turn the flow from the slots back upstream, with the result that if the slots are properly shaped and dimensioned, energy dissipation is excellent over a wide range of levels and discharges. It has the added advantages of permitting the fish to swim through the slots from one pool to the next at any desired depth, since the slot extends from top to bottom of the flume."

existing site as little as possible. The tests on the model had shown that more radical plans to remove portions of jutting rock on the east and west banks would only risk creating new and potentially damaging conditions. The fishways were experimental enough that a thorough reconstruction of the site was too risky to contemplate and, in any event, seemed unnecessary. As Edward Pretious later put it, "the scheme devised was to aid the natural river to perform its function, rather than substitute artificial features where the natural ones were adequate."[47] In the fall of 1944, with the support of the two national governments, construction crews began to excavate the site.[48] Built by Coast Construction Company under the supervision of Bell and the commission, the fishways cost over $1 million.[49] By 1945 one set of fishways was complete and the second was operable the following year.

A Justification and a Treatise

Despite Thompson's certainty of the causes of the seasonal blockages of salmon, his reasoned justification for the commission's building program did not appear in print until the fishways were complete. Published as the first bulletin of the IPSFC in 1945, his analysis of conditions at Hells Gate was a major statement on the history of salmon populations in the Fraser Basin and drew together a wealth of material developed over six years of commission research. Completed after Thompson had quit the commission in 1943 in frustration over political and personal disputes, the bulletin represented his personal commitment to and pride in the commission's scientific work.[50]

47. Pretious, "Salmon Catastrophe," p. 17.
48. The rationale for participation in the project was spelled out by Department of Fisheries staff in NAC, RG 23, Vol. 681, File 713-2-2[18], "Memorandum: Re: Permanent Fishway Facilities – Hell's Gate Canyon, Fraser River," February 22, 1944.
49. For a description of the building project, see "Preparing to Open Hell's Gate," *Pacific Fisherman* 43(1) (1945): 63. The total cost of fishways at Hell's Gate including later extensions was $1,351,000: International Pacific Salmon Fisheries Commission, "Hell's Gate Fishways," pamphlet, New Westminster, 1971, p. 5.
50. Thompson's resignation in 1943 followed on disputes with colleagues, particularly J. L. Kask, and continued problems with Tom Reid. Kask resigned in the same year after the blowup with Thompson. Thompson was also generally frustrated with the amount of time required for executive duties. UWA, Acc. 2597-3-83-21, Thompson, W. F. Papers,

Thompson cast the analysis of the problems at Hells Gate in a wide context. He offered a long-term explanation for shifts in the populations of Fraser sockeye, premised on the logic of racial analysis applied to historical data. The long view was enhanced by the specific knowledge of the timing of migrations and of the effect of obstructions gained through the Hells Gate investigations. Recent shifts in salmon populations were analyzed with a particular focus on the differential effects of Hells Gate on distinct racial units in the upper basin's various spawning grounds. As a whole, the analysis suggested a new race-based approach to future fisheries regulation and justified the construction of the Hells Gate fishways as the only reasonable way to restore Fraser sockeye to past levels.

Thompson's long view of the fishery pictured a healthy set of racial units buffeted by a series of significant and sometimes re-gionally specific environmental insults. Dividing the history of the fishery into five periods of decline and recovery since 1872, Thomp-son created a serial index of past sockeye populations based prin-cipally on catch records. He supplemented these data with other evidence and allowed statistically for changing rates of fishing pres-sure. Alongside the population index, Thompson examined chang-ing regional escapements as evidenced in spawning ground surveys and remaining hatchery records. This allowed for a specific analy-sis of racial units that had plummeted in years of decline or were responsible for general declines in the fishery four years later.[51] The first decline of the fishery, for example, he traced to the Quesnel Lake dam that existed from 1899 to 1903 without an operational fishway. Although Thompson allowed that overfishing might have added to the declines after 1903, he placed the primary emphasis on habitat destruction.[52]

Thompson identified a second major decline in salmon popula-tions also rooted in a specific environmental change: the building of the CNR through the Fraser Canyon beginning in 1911. This second event, however, had a broader impact across the basin and

Box 7, File, "Correspondence (re: Thompson's Resignation)," Thompson to A. L. Hager, Canadian Fishing Company, August 3, 1943 (copy). Thompson also complained about personal disputes and the politics of his position to his diary: Box 1, File "Diary 1943."

51. Methodological considerations are treated in Thompson, *Effect of the Obstruction*, pp. 22–39.

52. Ibid., pp. 50–5.

a longer-term, if variable, set of effects. Whereas the Quesnel Lake dam was specific to a number of racial units, the Hells Gate problem affected all racial units in the upper basin (and thus the vast majority of the Fraser sockeye population). But as Thompson and the commission scientists had discovered in the Hells Gate tagging experiments, the obstruction changed daily. At some water levels it blocked fish, at others times it provided passage. Thompson specified the consequences of this shifting impact by combining different environmental data: water level (recorded at Hells Gate since 1912, and extrapolated from Hope data for earlier periods), racial unit, and size of run (based on the latest data of typical migration dates and past spawner escapement information), as well as qualitative reports of regional population cycles. Viewed through the optic of racial analysis, these different strands combined to explain what had formerly appeared to observers like John Pease Babcock as wild upper-basin fluctuations.[53]

Consider the Adams River runs that had experienced a number of puzzling patterns in the two decades after the slides. Thompson charted the population history of the river's sockeye runs in relation to two key environmental events: the creation of a lumber splash dam in 1907 and the Hells Gate blockade of 1913 and after. The earliest impact was the easier to explain: A river blockage affected all Adams River runs, but was specific to that river because no parallel declines were experienced in other spawning grounds. The Hells Gate effect was more complicated. Just as different upper-basin runs experienced Hells Gate's variations differently, so too did the temporally distinct runs to the Adams River. The region received both early and late season runs of distinct racial units. In 1913, Thompson suggested, early runs survived, whereas later ones were diminished, some becoming extinct. Thereafter, problems remained, although they changed with seasonal water flows. Some runs experienced a precipitous decline, while others began to expand. In the course of ten years these shifts were registered in a transformation of the pattern of quadrennial dominance. Whereas before the slides, the 1913 cycle year was responsible for the greatest volume of spawners, after ten years the 1922 cycle year had

53. Ibid., pp. 84–156.

replaced it as the dominant run. In lay terms, this meant that salmon numbers peaked on a different four-year cycle than before 1913; Hells Gate was affecting the success and failure of upper-basin spawning runs by blocking some and favoring others. Overall the aggregate population had declined.[54]

This explanatory framework pointed to documented episodes of environmental destruction and explained their importance. It suggested why lower-basin stocks below Hells Gate had remained steady over the first three decades of the century while upper-basin runs fluctuated. But the analysis was closely tied to the tagging experiments. These studies supplied relatively precise data about how long it took specific races to complete their run to the spawning grounds in normal and delayed conditions, how resilient they were to delay, and how migration times were affected more or less than others. These experiments gave Thompson the confidence to state that the Hells Gate obstruction – and not overfishing – was the primary cause of the decades-long decline in Fraser River sockeye.

Conflict

But what if his assumptions were false? Thus did Bill Ricker put the question in a 1947 article in the *Journal of Wildlife Management* titled "Hell's Gate and the Sockeye."[55] Ricker was then a professor of zoology at Indiana University and a well-respected student of the sockeye and of West Coast fisheries. Holding a PhD from the University of Toronto, Ricker began his career at the Pacific Biological Station at Nanaimo and assisted Russell Foerster in his studies of sockeye at Cultus Lake. He had departed for Indiana in 1938 after working one season for the IPSFC at the time of the first discoveries of blockages at Hells Gate. It is unclear whether personal disputes had any role in his departure.

Starting from the position that Thompson's analysis required careful debate and scrutiny, Ricker leveled an empirical and interpretive critique of the Hells Gate study and raised serious doubts

54. Thompson, *Effect of the Obstruction*, pp. 20, 62–6.
55. Ricker, "Hell's Gate and the Sockeye."

about the necessity of the fishways. He started by focusing on a
key empirical finding: that during periods of blockage only 20%
of delayed sockeye were able to pass. This was an important point
because it underlay all of Thompson's claims about the rate of passage and the impact of delay on different racial units. The problem,
claimed Ricker, was that, although the figure reflected the data, the
data were so selective as to be unreliable and misleading. For one,
the sample taken at Hells Gate almost certainly did not represent
a cross section of the population, but likely contained a dispropor-
tionate representation of weak fish. Because strong fish could pass
the obstruction quickly, taggers would catch them less frequently
than they would weak fish. Moreover, the very process of tagging
intensified the weakness of the fish forming the major component
of the sample. Netting a fish, placing it in a box, clipping it, and returning it to water caused stress and sometimes split a fin – minutes
before fish were tested by the most difficult stretch of the river. Both
of these problems, Ricker stated, could have been accounted for by
more precise methods of data collection and by simple shifts in experimental design (changing the location of the tagging stations or
using control fish, for example). As it was, Ricker judged the short-
coming in the data to be important: "With regard to the possible
magnitudes of the effects of the above two sources of error, it can
be said without hesitation that they *may* be sufficient to completely
invalidate the conclusion that the Gate has been (1938–42) a serious obstacle to migration."[56] Change some of the assumptions
about the strength of the sample group, Ricker proposed, and it
may have been that the tagging sampled 80% of weak fish and
only 20% of the stronger migrants.

Thompson's findings were questionable in other ways. Why,
asked Ricker, was it plausible to assume that a correlation between problem water levels at the gate and spawner success in any
given year amounted to a cause-and-effect relationship? Climatic
conditions, after all, have variable effects across space. Although
high river flow levels might prove beneficial at the gate, they were
likely associated with flood conditions in upper-basin watersheds,
which would scour spawning grounds and reduce the success rate

56. Ibid., p. 13.

of the spawn. Water conditions at Hells Gate should not be considered as an independent variable, but placed within a wider context.

And there was no discussion in the report of the sex ratio of migrants past the gate. Given that it was widely understood that male spawners were more powerful swimmers than females, it logically followed, Ricker wrote, that a blockage would create a preponderance of male returns to the spawning grounds. Spawning ground surveys in the years of blockage, however, provided no evidence of abnormal sex ratios. Did this mean that the appearance of a block at the gate was false? Possibly, Ricker said; at least it required explanation.

What, then, was one to make of the fishways? If the proof of blockage conditions was in doubt, so too was the necessity for this expensive conservation measure. In the absence of other conservation measures, if upper-basin spawning grounds were rebuilt, then, Ricker judged, the fishways would surely be deemed worthwhile. He worried, however, that they were more likely to serve as an excellent excuse to avoid problems of overfishing. Although Thompson's report discounted fishing as a primary cause of declines, he did report, Ricker underlined, that "the commercial fishery may take about 80 per cent of the sockeye returning from the sea; and tag returns show that 50 per cent is the absolute minimum."[57] What if the fishways were not about to save sockeye spawners? Would it not be worth considering stringent catch controls, at least to enhance the fishways' possible success? It would be a "gamble," Ricker concluded, to leave the task of conservation to only the fishways.[58]

Ricker's paper was framed as a scientific critique of an admirably complex study. It did not, however, shy away from drawing strong conclusions about the wisdom of the IPSFC's research and building program as well as questioning William Thompson's capabilities as a scientist. If it was intended as a disinterested critique, it was not received in that spirit. The paper led to a major scientific controversy in the fisheries research community that spilled into the fisheries press and was cast by its participants along national lines. The

57. Ibid., p. 19.   58. Ibid.

international cooperation inherent in the IPSFC program seemed for a time in tatters.

The depth of feeling that Ricker's paper aroused is revealed in the correspondence between Thompson and some of his closest colleagues in the fisheries research community. Days after the journal was printed, Richard Van Cleve, the IPSFC chief biologist and a professor in the College of Fisheries (UW), registered his dismay to Thompson.[59] Van Cleve did not comment on the scientific aspects of Ricker's paper, but judged it as the expression of a "personal grudge" against Thompson and the IPSFC, although with wider implications. Van Cleve argued that Ricker's article was, "in effect an attack on all biological fisheries work on the Pacific coast and will result in casting a doubt on the validity of any of our work, especially that on salmon." To counter this effect, he urged Thompson to respond with an accessible piece that would win over a general audience.[60] Thompson appeared to agree with Van Cleve's reading of events. Writing to Fred Foster, formerly the regional director of the U.S. Bureau of Fisheries in the Pacific Northwest, he explained that the controversy was more political than scientific. Ricker was formerly a member of the Biological Board of Canada, he explained. The board had not discovered the problems at Hells Gate; its policies were abandoned in light of the IPSFC's work and, as a result, were made to look ill advised. Ricker, he thought, was salvaging the reputation of the past Board and its research. "These Canadians," he wrote, "are somewhat in the position of a man who sat on a powder keg while the fuse burned, telling the world that it could not blow up."[61] Already improved returns through the fishways in 1946 were showing that the commission had been right. It was his duty, Thompson explained, to air the debate for what it was.

Thompson's subsequent response to Ricker's paper shifted the controversy from Hells Gate to the credibility of Canadian fisheries science. Rather than focusing on Ricker's published criticisms, he

59. UWA, Acc. 1683-71-10, Van Cleve, Richard Papers, Box 4, File Correspondence to Thompson, 1947, Van Cleve to Thompson, January 21, 1947 (copy).
60. Van Cleve later read a preliminary version of Thompson's reply: PSCA, File 1180.1–74, Van Cleve to Thompson, May 5, 1947 (copy).
61. UWA, Acc. 2597-77-1, Thompson, W. F. Papers, Box 15, File 29, Thompson to Fred Foster, February 14, 1947.

reviewed the history of research on the Fraser sockeye and judged it wanting. Even the work carried out at Cultus Lake that had been widely hailed as the most exacting examination of the efficiency of artificial propagation (and in which Ricker had had a hand) was cast in the same light: "None of these investigations led to positive remedial action, successful or otherwise." The Hells Gate situation, meanwhile, went unstudied. The Fisheries Research Board of Canada (FRBC) he claimed, "either tacitly, or actually, acquiesced" in the "official view" that nothing was amiss at Hells Gate after the initial cleanup. "Either the problems at Hell's Gate were not appreciated by the Research Board and Dr. Ricker, or as often may happen in governmental work, an 'official' view was allowed to modify the research program, consequently its results." Either possibility was a stinging indictment of Canadian scientists: In this representation they were fools, or lackeys, or both. Thompson reserved some space to attempt to dismantle each of Ricker's critical arguments, but much of his defense rested on spawning returns after the construction of the fishways. Fish numbers were improving; therefore the fishways were necessary and a success.[62] The reply was mimeographed and sent to over fifty scientists in the United States and Canada, to the main fisheries dailies, to the IPSFC commissioners, and to select politicians.[63]

Two fisheries journals featured the reply and spun out the story as a significant battle among national fisheries science communities. *The Pacific Fisherman* defended Thompson's position entirely. The editor of the journal showed a preliminary version of the story to Thompson and heaped scorn on Ricker,[64] cast as a "scientific sharpshooter."[65] In a subsequent article, aiming to provide equal space to the opposition, the editor derided Canadian scientists as defensive and evasive.[66] He said that Ricker's role was personally

62. BCARS, GR 1378, BC Commercial Fisheries Branch, Box 3, File 5, Thompson, "Hell's Gate Blockade and Salmon," March, 1947.
63. UWA, Acc. 2597-77-1, Thompson, W. F. Papers, Box 15, File 29, B. M. Brennan, director of IPSFC, to Thompson, April 1, 1947. Brennan's letter lists forty-seven individuals and institutions to whom Thompson's paper was sent and included twenty-five more reprints for Thompson to send personally.
64. UWA, Acc. 2597-3-83-21, Thompson, W. F. Papers, Box 8, File, "Ricker's criticism," Stedman H. Gray, executive editor, *Pacific Fisherman*, to Thompson, April 10, 1947.
65. "Scientific Sharpshooting," *Pacific Fisherman* 45(5) (1947): 37.
66. "Do Nothing Biology," *Pacific Fisherman* 45(7) (1947): 30.

motivated and political. The release of Ricker's article shortly before Canadian parliamentarians were to reassess IPSFC funding was said to be "significant" and deliberately destructive. Like all of Ricker's critics, the journal stated, "The proof of fishways is in the fish which pass them."[67] *The Canadian Fishermen's Weekly*, by contrast, seemed to side with Ricker initially, or at least to give him a platform.[68] Subsequently the journal published a filtered conversation between Ricker and Thompson, as they sparred back and forth in public view. The journal also reported the views of commissioners, such as Tom Reid, who lashed out publicly against Ricker's statements and allowed members of the FRBC the opportunity to defend their research record.[69]

Members of the board were personally affronted by Thompson's public remarks and conducted a campaign to defend the reputation of their institution and themselves. Wilber Clemens, who had been director of the Pacific Biological Station in the period of alleged negligent research, prepared his own mimeographed response for wide circulation. In it, he reviewed the research projects of the decades before the commission came into being, and pointed out that none of them aimed specifically at rehabilitation activities, as Thompson had suggested. To fault life-history research for not turning up the Hells Gate problem was misleading and unfair, he charged. More to the point, Canadian research had been held up because of the interminable delays in ratifying the Pacific Salmon Convention[70]:

The Fisheries Research Board was not asked to undertake a general investigation of the Fraser River with the objective of rehabilitating the sockeye runs because from the time of the establishment of the International Fisheries Commission (Halibut) in 1923, negotiations were almost steadily in

---

67. "Scientific Sharpshooting," p. 30.
68. "Hell's Gate and the Sockeye," *Commercial Fishermen's Weekly* XIII(8) (March 14, 1947): 90–1.
69. The key articles are "Salmon Commission Hits Back at Critic," *Commercial Fishermen's Weekly* XIII(10) (March 28, 1947): 111, 113; "Review of Evidence Suggested by Ricker," XIII(12) (April 18, 1947): 135–7; "Research Board Said Not Open to Charges," XIII (13) (April 25, 1947): 152–3.
70. BCARS, GR 1378, BC Commercial Fisheries Branch, Box 3, File 5, W. A. Clemens, "A Statement Regarding the Memorandum 'The Hell's Gate Blockade and the Salmon,' by W. F. Thompson," April, 1947.

progress for the establishment of an International Commission for dealing with the sockeye salmon problem of the Fraser River.

In personal letters both Wilber Clemens and Russell Foerster criticized Thompson for drawing the FRBC into the debate.[71] Foerster described Thompson's views as "totally incorrect and misleading."[72] Thompson responded by standing by his remarks and pointing out that the many activities carried out by the Biological Board in the years before the IPSFC had done little for the rehabilitation of Fraser sockeye. He had heard that Ricker had aired his views to Canadian scientists before publication and that Ricker had been encouraged to proceed. Why, Thompson asked, had he or the commission not been notified in advance before such destructive criticism was unleashed? Thompson stated plainly that he would not stop criticizing Ricker until his point was understood: "There are deeper issues at stake than mere argument."[73] In 1948 the executive of the FRBC passed a special resolution in the proceedings of its annual meeting, condemning the IPSFC for Thompson's criticisms of the FRBC's past research and calling on Canadian commissioners to state publicly whether they agreed with the allegations. Dr. Dymond, a distinguished University of Toronto scientist, sponsored the item and Wilber Clemens seconded it.[74]

The actions of the FRBC executive in calling on Canadian commissioners to dissociate themselves from Thompson's views suggest something of the complex national and international politics that occurred in the Hells Gate debate. Whereas before the controversy, the most obvious signs of national antagonism occurred within the IPSFC, after Ricker's critique a remarkable solidarity developed within the commission against the perceived external threat. Tom Reid, for example, a frequent critic of Thompson within the commission and a reputed cause of Thompson's departure from the directorship of scientific investigations, defended Thompson's work in his position as commission chairman. Ironically, Ricker's article

71. UWA, Acc. 2597-77-1, Thompson, W. F. Papers, Clemens to Thompson, April 8, 1947.
72. UWA, Acc. 2597-3-21-83, Thompson, W. F. Papers, Foerster to Thompson, April 10, 1947.
73. UWA, Acc. 2597-3-21-83, Thompson, W. F. Papers, Thompson to Clemens, May 8, 1947 (copy).
74. NAC, RG 23, Box 682, File 713-2-2[26], Fisheries Research Board, Extract from an Executive Minute, June 9–11, 1948, Vancouver, BC.

and Thompson's criticisms of the FRBC and Canadian science had the effect of lessening national differences within the IPSFC. Such a drawing together would increase in the 1950s when the threat of dam building created another common cause.

There were no doubt personal, national, and scientific aspects to this debate. Thompson was bitter over the personal and political conflicts within the commission, particularly those involving Jack Kask and Tom Reid. Although the nature of these disagreements is unclear, their depth is not: In 1943 both Thompson and Kask resigned, citing their poor relationship as a key reason.[75] Thompson's certainty that Ricker's critique was primarily a grudge was born of the paranoia he had developed while operating in a politicized scientific environment. Although it would appear that Ricker's motivations were more properly scientific than Thompson allowed, he also intensified the controversy by using provocative statements in his paper and to the press. Personal and national antagonisms seemed to share some common ground. Thompson's relationships with each of the Canadian scientists originally hired in 1938 had soured by the time of the controversy. At a more fundamental level, his low opinion of past Canadian fisheries research reflected a divergence in national styles of fisheries management. Whereas after 1935 Canada opted not to use hatcheries as a management tool, following studies that suggested their negligible effect, in the United States their importance only grew.[76] Here lay the basis for Thompson's disparaging comment about the poor remedial work of Canadians.

These personal and national tensions ensured that the underlying scientific issues in the debate were overshadowed by the perceived motivations of its participants. Tim Smith, a fisheries scientist and historian, claims that at the heart of the debate was a fundamental disagreement about the role of overfishing in fisheries depletion.[77] Yet Thompson was so bent on defending his reputation that he rarely engaged with Ricker's point that the IPSFC's conservation program was primarily aimed at restoring habitat rather than

75. For a biography of Kask and mention of the dispute with Thompson, see Johnstone, *The Aquatic Explorers*, pp. 208–9.
76. Taylor, "The Political Economy of Fishery Science."
77. Smith, *Scaling Fisheries*, pp. 276–85.

controlling fishing. Nor did he launch a detailed justification for his belief that fishing pressure exerted a much less serious effect on fish populations than others claimed. After Ricker's contention was dismissed, the debate became political rather than scientific and did not focus on these key questions. In future research, Ricker would develop what became known as the spawner-recruit theory to establish the effects of spawner success on fry development.[78] His concerns were not merely a reaction to the fishways project or to Thompson's bulletin, but were part of a longer-term consideration of the limits of fisheries.

Resentment over the allegations and counterallegations in this debate lasted for years in the BC fisheries science community.[79] But the controversy did not delegitimize fisheries science as a whole as Richard Van Cleve had feared. The public perception of the Hells Gate research program was, to the contrary, almost entirely positive. The public favored the idea of restoring the Hells Gate site once and for all; moreover, salmon populations had increased.

The fishways were greeted publicly as a miraculous exercise in technical mastery over nature. Completed in the euphoria of war's end, journalists described the fishways as one more battle won, a great public works project linking coast and interior. Hells Gate was a door unlocked and pushed ajar, a barrier overcome with a highway, a staging ground for the "invasion" forces of salmon. Scientists were miracle workers with keys, "tough men," freedom fighters.[80] The connections in these representations between the commission scientists and armed struggle suggest not only the saturation of military metaphor in public discourse, but also the ideas of science as liberator and scientists as hard-working soldiers. In praising the Hells Gate studies in 1942, a *Vancouver Sun* editorial described the scientists' work as "definite and clear, completely proven – checked a score of times to prevent the possibility of

78. Ibid., pp 285–292.   79. Roos, *Restoring Fraser River Salmon*, p. 306.

80. Clippings were found in NAC, RG 23, Box 682, File 713-2-2[23], "Fish Travel Modern Highway," *Province*, October 28, 1944; "Hell's Gate Soon Ajar for Salmon," *Western Business and Industry*, January 1945, Vol. 19(10), pp. 8–9; "Ready for Salmon Invasion," *Province*, June 2, 1945; *Vancouver Sun*, August 6, 1945, cartoon; "Hell's Gate a Job for Tough Men," *Province*, August 11, 1945; "Hell's Gate Unlocked by Science," *Seattle Times*, September 30, 1945; "Freeing of 'Hell's Gate'," *Ottawa Citizen*, January 28, 1946; "Hell's Gate Aids 'Miracle' of Nature," *Province*, October 2, 1946.

Figure 2. "Fish Travel Modern Highway."
(*Source:* NAC, RG 23, Box 682, File 713-2-2[23], "Fish Travel Modern Highway," *Province*, October 28, 1944 [?], by Ray Tracy [?].)

error."[81] The virtues of an idealized science became the virtues of the fishways. In the ultimate representation of the transformation of Hells Gate from turbulent passage to domesticated space, one cartoonist drew passing salmon as ordinary citizens involved in a commute (see Figure 2). The fishway was a modern transportation system. Salmon passing through were dressed in the attire of business and lay people. "I've been herring a lot about this fishway!" declared one, toting a briefcase. "Let's rest behind the next baffle," said another. Beside this kind of public enthusiasm, the debate between Thompson and Ricker had little broader resonance.

It is also important to note the extent to which the fishways appeared to be working. G. B. Talbot's study of the efficiency of the fishways in passing fish at problem water levels judged it to be high. Using the same tagging methods to gauge the passage of fish as were used in the original experiments, Talbot found that the fishways eliminated the periods of seasonal delay that had played such an important role, in Thompson's view, in diminishing the capacity of salmon to spawn successfully. Furthermore, the commission's counts of returning spawners, the so-called escapement figure, showed a marked increase following the final completion of fishways in 1946. "After installation of the fishways," Talbot summarized, "the mortality rate between Hell's Gate and the spawning grounds was reduced approximately 20 per cent to 30 per cent."[82] Notwithstanding other factors (and there were many), the fishways

81. "Ottawa Cover Up on Bygone Errors," *Vancouver Sun*, February 9, 1942.
82. Talbot, *A Biological Study*, p. 77.

appeared to have provided the basis for a rise in Fraser sockeye populations in the postwar period. Commission scientists spoke publicly of the possibility of restoring hitherto forgotten and depleted runs. Although the enhanced regulations of the fisheries played an important role in this expansion of sockeye populations, commission scientists marshaled significant data to suggest that obstructions were much less serious than in the past. Fishways were added in the late 1940s at Hells Gate, the Bridge River Rapids, and Farewell Canyon.

Conclusion

Hells Gate haunted fisheries scientists, regulators, and native and commercial fishers for decades. For years questions surfaced as to whether the gate was cleared. One of the major proponents of the Pacific Salmon Convention, John Pease Babcock, consistently argued that the problem was solved. A convention was needed, he argued, to control fishing, the real culprit of fisheries depletion. Ironically, the scientific investigations carried out under the auspices of the commission found the opposite to be the case.

The investigations centering at Hells Gate under William Thompson's leadership operated within a natural–cultural nexus. Scientific data did not simply represent nature; it was created by methods that produced various forms of natural and cultural selection. The confusion of scientists collecting tags from native fishers followed and reproduced established patterns of interaction among fisheries officials and natives in the canyon. The identification of problems at Hells Gate had rippling effects in local communities, as celluloid disks became currency equivalents – with effects on the final data that are impossible to know.

No single environmental event was as important in fixing the judgment of Thompson and his research team as the water conditions in the canyon in 1941. The spectacle of 6 miles of mature sockeye turning red below the gate convinced Thompson that Hells Gate was an obstruction that must be cleared. Through the prism of this event and the data collected in the tagging experiments, he analyzed the history of sockeye populations in the basin as a saga of fish and dams. Correlations between an index of population size and environmental insults demonstrated, in his view, that the primary causes of fisheries decline were to be found in episodes of habitat destruction. A healthy fishery needed clear passage.

The fishways constructed in 1945–6 to bypass turbulence at Hells Gate were said to restore the river to its natural condition. Rather, artifice had been placed upon artifice. An unnatural dam had been deposited in the gorge, and the fishways were an unnatural response. Science tamed the gorge, as the newspapers never failed to suggest, and made a rough passage into a salmon highway. Subsequent studies argued that the fishways facilitated significant expansions in upper-basin spawning runs in the postwar period.

The remaking of Hells Gate, however, raised various questions, some scientific, some national, and others personal. Bill Ricker asked whether the IPSFC data could be trusted. William Thompson replied that Canadian scientists were carping after realizing their own errors. The dispute tested the collegiality of the fisheries science community after years of national tension within the IPSFC; it also strengthened the internal coherence of the IPSFC. The science of Hells Gate remade careers, reputations, and institutions as well as water and fish.

The undamming of the gate had lessons for scientists, politicians, and the fishing industry about the dangers of dams, lessons that required increased prominence in the late 1940s. The counterexample of the Columbia, with three major main-stem dams by 1941, modeled the dangers, but also produced the knowledge to create the fishways. Proposals in BC to dam the Fraser gained credence by the end of the war and were proposed by a variety of private interests. Hells Gate could stand as a monument against these proposals, but it also raised expectations: If scientists could tame this beast, why not another human-designed dam, where fishways would be integral to the design? The enigma of Hells Gate was reproduced in the fish vs. power debate of the following decade. Its meaning could not be fixed.

In a parting salvo in his critique of the commission's science, Bill Ricker had raised an intriguing idea about the possible rationale for the fishways. Maybe, he mused, the IPSFC wanted a fishway to ensure that the Hells Gate site, and the canyon around it, would be safe from hydroelectric-power developers. For once the fishways were built, publicity created, and salmon apparently saved, who then would think it permissible to sink this binational investment under the placid waters of a reservoir?[83] There seems little basis for this suspicion, but Ricker was not alone in the direction of his thinking. BC Water Branch officials fretted that the fishways would destroy their plans for major postwar water developments in the canyon.[84] The fishways not only saved fish, they also claimed territory.

83. Ricker, "Hell's Gate and the Sockeye," p. 19.
84. BC Water Management Branch, Department of Lands 'O' Files, File 5254, Davis to minister of lands, April 11, 1942; BCARS, GR 1378, BC Commercial fisheries Branch, Box 3, File 3, George Alexander to commissioner of fisheries, May 30, 1942 (copy); E. Davis, comptroller of water rights to Alexander, July 7, 1942; deputy attorney general to Alexander, July 16, 1942; H. Carthcare, deputy minister of lands to Alexander, July 17, 1942; Alexander to commissioner of fisheries, July 20, 1942 (copy). Alexander's correspondence here cited pointed to his irritation with the IPSFC for not advising the provincial commissioner of Fisheries. Other items discuss how best to counter or condition the fishways application.

4

Pent-Up Energy

If one were to choose any particular moment in the years after World War II in which the hopes, self-doubts, and politics of power were on display in full dress, none could serve as well as the inauguration ceremony for the BC Electric's Bridge River project in 1948. Taking place just months after a major flood displaced tens of thousands of British Columbians, the event was a cathartic experience of self-affirmation in which the virtues of electrical technology were praised and the divisions it created in society downplayed. The ceremony mixed tradition and modernity, private enterprise and public sanction; it acknowledged past shortcomings and pointed to their present rectification.

The symbolism paraded at the Bridge River powerhouse on October 24, 1948, bespoke a new, electrified BC. In front of 200 Vancouver business people and municipal and provincial politicians, Dean Cecil Swanson of Vancouver's Christ Church Cathedral dedicated the project to "the Glory of God and the service of man." Shuffling to the podium, an aged Geoffrey Downton, the first surveyor to identify the Bridge River site in 1912, sounded the official siren to open the penstocks and let the waterpower flow. He envisioned that the project would "brighten the lives and lighten the toil of countless thousands in the years to come." Acting Premier Herbert Anscomb told members of the press that the Bridge River project was a marvellous addition to the BC Electric's "great free system." If another world war came, he intoned, the project would gain even greater significance. Editorial writers noted that BC Electric would now be selling excess power to the Bonneville Power Administration (BPA) in the United States along power lines that only a year before had served as an electrical lifeline in the opposite direction. What had once been a cause for shame was now a source of pride. BC, the editors implied, had come into its own.[1]

This was the beginning of a new hydroelectric era in BC. After no substantial expansion to the province's electrical supply during the 1930s and sharply increasing demands during the war, the late 1940s witnessed a host of new power-development and transmission projects. BC Electric expanded its power supply for its urban markets. The newly invented Public Power Commission consolidated the hinterland market, developed a rural electrification program, and began to build dams on Vancouver Island and in the Okanagan. By the end of the war a distinctive, mixed system of private power in the cities and public power in the hinterland had emerged. The spirit of the times was conveyed well by one nervous salmon official, who observed in 1947 that "If you shake a tree another engineer falls out."[2]

Wartime dissatisfaction drove change. Criticisms of the relatively poor state of rural electrification, the high rates for electricity, and the limited electrical supply led numerous groups and commentators to call for a grand solution: state expropriation of the electricity business. Where private industry had failed, went the hope, public power would create the basis of a new future, in city and hinterland alike. In part these concerns emerged from pragmatic attempts to extend electrical service with expediency; in part they connected with a more general shift in politics that saw a new role for the state in economic development and collective affairs. At a time when the traditional parties of the center and right in the wartime coalition government sought to diminish the rising popularity of the Socialist Left under the Canadian Commonwealth Federation (CCF), the public power issue became a contentious battle ground

---

1. BCER CF, "Bridge River Just in Time to Avoid Power Brown-Out," *Province*, November 25, 1948; "In a Position to Help," *Victoria Daily Times*, November 22, 1948; "Vancouver Seen as Industry Hub, 'New York of Canada,'" *Vancouver Sun*, October 28, 1948; "Former Victoria Man Honored in Opening of Electric Plant," *Victoria Daily Times*, October 25, 1948; "Prayers, High Hopes Dedicate Power Dam," *Province*, October 25, 1948; "Bridge River Development Work Initiated Many Years Ago," *Vancouver News Herald*, October 23, 1948; "Power Line Defeats Mountains," *Province*, August 28, 1948. The quotations are from the *Victoria Daily Times* article of October 25, 1948.
2. NAC, Pacific Region, RG 23, Vol. 2301, Folder 6, Proceedings of the IPSFC meeting, August 9–10, 1947.

in the shaping of a postwar political economy. Against the backdrop of the early Cold War, decisions about the organization of power development signaled broader choices and concerns.

In the midst of the postwar building phase, the Fraser River flooded. In the spring and summer of 1948, the river inundated the Fraser Valley, severed major transportation routes, and threw the province into a state of anxiety. As the flood receded, the federal and provincial governments agreed to conduct a joint river management strategy with a view to developing flood control dams. Thus, added to the hydroelectric building program would be a distinct, but related, concern for flood control.

The consequences of the power scramble and the flood shaped BC river politics into the 1960s. This chapter seeks to explain the origins of institutions, the social impetus to development, the politics of power, and the making of a new era in river management policies. The aim, in short, is to peel back the layers of the Bridge River inauguration, to peer at the contradictions hidden in ceremony, and to listen for the conspicuous silences.

Imagining Postwar Power

In 1942, members of the Duncan Rotary Club sat down to write a letter to the Post-War Rehabilitation Council. So did the Summerland Women's Institute and the Prince George Junior Chamber of Commerce. Besides their common participation in the council's still premature dreaming of a postwar future, these groups and others like them also had a similar idea. The government, they wrote, would do well to expand electrical power in the province, make it available in isolated areas, and put idle rivers, like the Fraser, to work.[3]

The council dealt in such matters. Appointed by the provincial government in 1942 to plan for the inevitable dislocation of

3. Different organizations and individuals that made submissions on the topic of electric power, hydroelectricity, and waterpower are listed in the Appendix to Hon. H. G. T. Perry, *Interim Report of the Post-War Rehabilitation Council*, p. 22. Some short versions of proposals offered in these submissions are listed in a later section of the Appendix, pp. 57–9, 83–5. It is evident that some groups listed in the second section are not listed in the first. I have combined the two lists to arrive at a rough total of thirty-eight.

the postwar period, a team of ten Members of the Legislative Assembly (MLAs), primarily from the Coalition government and led by education minister and former mayor of Prince George, Harry G. T. Perry, toured the province that year and solicited the views of individuals and groups by mail.[4] In their number was a handful of politicians who would have an important impact on BC's postwar political scene and hydroelectric development: W. A. C. Bennett, the future Social Credit premier, but at this date a Conservative, would oversee the creation of BC Hydro and major developments on the Columbia and Peace Rivers; E. T. Kenney, a Liberal and future minister of lands and forests, would shepherd the development of a large dam and aluminum smelter project in the late 1940s; and Harold E. Winch, leader of the provincial CCF, would remain a staunch advocate of public power throughout the postwar era. The Post-War Rehabilitation Council not only focused the attention of the province on postwar possibilities, but also opened the province and its regions to this group of politicians as never before.[5]

In reply to the written solicitations of this band of traveling politicians, British Columbians, it turned out, had much to say. Tourism, forestry, fisheries, and road building attracted comment and recipes for change. In their report, the councillors weighed these views, summarized some of them, and published a few compelling briefs in full; they also chose to defend key ideas as their own. On the issue of hydroelectric development, and in contradiction to past Liberal and Conservative policy in BC, the report came down firmly in favor of a postwar role for the state in hydroelectric development and electrical distribution. Electricity, the report argued, had become a necessity of life; its reliable supply was a foundation of society. State intervention in this sector could help to reduce rates,

4. Members of the Council were as follows: H. G. T. Perry, W. A. C. Bennett, E. T. Kenney, C. G. MacNeil, J. A. Paton, W. T. Straith, H. E. Winch, Mrs. N. Hodges, Mrs. T. J. Rolston, and Mrs. D. G. Steeves. The latter three female councillors were added late to the council, perhaps to make an attempt at bridging the gender imbalance. Perry's *Interim Report* stated "that letters and copies of the Post-War Rehabilitation Act were sent to all public bodies in the Province. These included: – Cities, District and Village Municipalities, Boards of Trade, Chambers of Commerce, Veterans, Farmers, Labour, Manufacturing, Industrial, Trade and Service Organizations, Women's Institutes and numerous other public organizations" (p. 9).

5. Mitchell, *WAC Bennett*, pp. 72–4, 76, 303.

provide better provisions for emergencies, serve new areas through interconnections, develop more waterpower, and decentralize industry.[6]

In the Appendix, a forceful brief amplified these points and stressed the need for immediate action.[7] Written by Harry Warren, a UBC geologist who would come to play a prominent role in BC power politics a decade later, the document attempted to point to past failings and suggest the best means of rectifying them before the end of the war. Warren started by reviewing BC's potential energy sources and then listed the relatively small amount of developed hydroelectric power. He mused about the reasons why power had not been more actively developed in the past and decided that the high capital cost and the lack of awareness of BC's potential in other parts of Canada and the world were to blame. Warren's estimate of the necessary funding ran to $90 million, with an annual return on developed power of $10 million. To gain a sense of the costs, promote the province, and involve the government in these undertakings, no time could be lost. Waiting until the end of the war would be "entirely too late," he insisted.

Where were these great development opportunities? Many of the briefs gestured vaguely toward the Fraser or mentioned one of its tributaries. These views reflected the fact that the majority of the population lived in the Fraser Basin and considered the river's development as the most feasible and propitious. Provincial government staff, in charge of surveying BC's waterpowers, made parallel observations. Ernest Davis, the provincial water comptroller, stated to the council that the best postwar water-development opportunities existed in the Fraser Basin.[8] Three exceptional sites for diversion of interior flows to the coast were located in the upper basin; in

6. Perry, *Interim Report*, p. 131.
7. Harry V. Warren, "Excerpts from a Brief Submitted to the Post-War Rehabilitation Council," in Appendix of ibid., pp. 419–21. Warren also made such views pubic in a speech to the Vancouver Board of Trade in 1942: BCER CF, "Dr Warren Urges Hydro Development," *Vancouver Sun*, October 16, 1942; "Raw Material Wealth Answer to Postwar Problems," *Province*, October 16, 1942.
8. BCARS, GR 1006, BC Water Rights Branch, Box 1, File 10, comptroller of water rights to Hon. H. G. T. Perry, chairman, Postwar Rehabilitation Council, September 8, 1942 (copy). Davis also promoted the power possibilities of the Fraser during the war in the business press, Ernest Davis, "Fraser Drainage System Could Furnish Power for Giant Metallurgical and Chemical Industries," *The Financial News*, October 31, 1941, and "Development of Water Power in British Columbia," *British Columbia Financial Times*,

addition, the main stem afforded major possibilities. Interestingly, the key sites of postwar development on the Columbia and Peace Rivers gained barely a mention: The Columbia's importance was described as modest, and the Peace was not listed at all. Here, then, was a telling register of midwar attitudes of the geographical boundaries of power development. The limitations of transmission costs, of course, informed Davis's advice. But one also suspects that distant developments, outside of the core region of the province, could not quite be imagined at this date.

Universal Electrification?

Governments shelve undesirable reports. The Post-War Rehabilitation Council's firm advocacy of public power did not lead to this result. Instead, following the council's recommendation to study the issue, the Coalition Cabinet established a committee of civil servants for this purpose in 1943 with instructions to focus on rural electrification. The mandate, admittedly, narrowed the focus, but the possibility of state intervention opened as never before.

The emphasis on rural electrification reflected the concerns of many of the participants in the hearings of the Post-War Rehabilitation Council. Although Warren believed in the possibility of an integrated, developed electrical state, most of the submissions to the council asked simply for local electrical hookups or a drop in electricity rates. Of the thirty-eight submissions to the council on the topic of electrical development, thirty-one of them originated outside of Vancouver and Victoria.[9] Although civic populism had been a major political force in pressing for public power in earlier decades, urban groups were conspicuous in their absence from the roll call amassed by the council.

Rural electrification as an ideal had broad appeal in the war years. During the New Deal era, the American government had developed a number of state-funded projects to expand the

XXVIII(21), November 1, 1941. These articles are contained in the same file as Davis's brief.
9. Of the seven submissions I identify as originating in Victoria or Vancouver, provincial government bureaucrats submitted two.

boundaries of the electrified universe. The TVA, the Columbia River projects, and the Rural Electrification Administration (REA) all received wide publicity in the late 1930s.[10] Pioneering examples in Canada, such as the Ontario Hydro-Electric Commission, were also the point of frequent comparison.[11] Across Canada, from around 1942, rural electrification emerged as a major subject of postwar planning.[12] The growing appeal of collectivism during the war, the enhanced role of the state, and the rise of the organized left all forced the agenda.

Rural perhaps was a misnomer, at least in BC.[13] The object of consideration was nonmetropolitan BC: the province, less Vancouver and Victoria. Unlike the New Deal REA program, the mandate of the Rural Electrification Committee (REC) focused not on dispersed settlements engaged in agriculture, but on hinterland regions of the province. These included smaller cities in the urban hierarchy, resource towns, and regional service centers as well as dispersed settlements and farms. Such areas received electrical services from a host of small systems owned in some cases by municipalities, run as offshoots to industrial projects or as minor private corporations. One dominant firm, on the other hand, BC Electric, controlled the urban market in Vancouver and Victoria, as well as the interior city of Kamloops (see Table 2). Although the perception existed that the hinterland regions were poorly served and paid dearly for electricity relative to the cities and other parts of the country, arguably the entire province experienced relatively high rates. The committee's comparison of rates in different BC locales with communities and cities of similar size in Ontario showed in general that

10. Nye, *Electrifying America*, Chapter 7, "Rural Lines," pp. 287–338; Tobey, *Technology as Freedom*.
11. Although a number of provinces had experimented with rural electrification programs before the war, Ontario had by far the greatest level of rural extension: Fleming, *Power at Cost*, p. 16. The interest in the experience of other jurisdictions is well represented in the discussion devoted to the subject in Perry's *Interim Report*.
12. In 1945 the *Vancouver News Herald* presented results from a survey by the Canadian Press that showed that every province had some form of postwar rural extension program in development: BCER CF, "Rural Electrification Interests Every Province," *Vancouver News Herald*, February 28, 1945.
13. The committee dispensed with the terms rural and urban and referred to different systems by numbers of consumers, classified into five groups. For a discussion of the term rural as an analytical category and a conceptual boundary marker, see Sandwell, "Finding Rural British Columbia."

Table 2. *Central Station Groups, 1942*[14]

Central Station Group	Number of Consumers	% of Total Provincial Consumers	kWh Production	% of Total Provincial Production
BCER group	152,762	73.90	634,268,540	85.35
West Kootenay Power and Light	8,289	4.01	42,234,446	5.68
Northern BC Power Co.	3,011	1.46	16,636,480	2.24
West Canadian Hydro-Electric Corp.	4,390	2.12	9,524,001	1.28
Nanaimo–Duncan Utilities	6,129	2.96	3,760,063	0.51
Five companies subtotal	174,581	84.45	706,423,530	95.06
Twenty-six other private utilities	4,941	2.39	5,657,694	0.76
Nineteen municipally owned utilities	21,907	10.60	8,754,834	1.18
Fifteen industries and institutions	5,294	2.56	22,258,436	3.0
Sixty-five total for province	206,723	100	743,094,494	100

Source: Rural Electrification Committee, *Progress Report*, Table 13, p. 52.

British Columbians used less than half the domestic electricity of Ontarians while paying more than twice the cost per unit of power (see Table 3). The problem, then, was how to extend electrification to the remaining undeveloped regions while making electrical use more accessible and affordable in areas of existing, but unsatisfactory, service.

The word rural also signaled a set of political concerns. By 1943, the coalition government had produced a number of blunt signals that it intended to expropriate private utilities after the war: The Post-War Rehabilitation Council's report explicitly advised this route. In the summer of 1943, Premier John Hart issued a press release that appeared to leave little doubt as to his government's

14. This table summarizes information from the Rural Electrification Committee's *Progress Report*, 1944, Table 13, p. 52.

Table 3. *Comparative Electrical Costs and Domestic Consumption for Systems of Equal Size in BC and Ontario, 1942*[15]

Distribution System	No. of Customers	Avg. Annual Domestic Consumption (kWh)	Avg. Charge per kWh (Domestic) in Cents
Lower Mainland[a]	105,507	1,068	2.45
Toronto	178,956	2,400	1.15
Victoria and Region	22,332	895	3.12
London	21,373	2,952	1.03
Nanaimo	4,085	609	5.4
Brockville	3,491	1,884	1.10
Kimberley	1,479	1,098	3.5
Bowmanville	1,381	1,716	1.58
Courtenay	940	483	6.2
Prescott	959	2,196	1.28
Princeton	570	530	4.3
Caledon	557	804	1.9
Quesnel	252	368	9.7
Elmvale	251	912	2.2

[a] Not including the Fraser Valley.
Source: Rural Electrification Committee, *Progress Report*, 1944, pp. 28–31.

direction: "The proposal," Hart stated, "is for the Province to take over development of power and furnish it to municipalities at arranged centres. The Government thus will participate in the purchase to this extent."[16] Despite pointing to the direction of policy, however, this statement was unclear as to how such a transfer would be arranged. The ambiguity may have been deliberate: From one perspective, the coalition government's flirtation with the issue of public power was a calculated attempt to capture nonmetropolitan support and keep the rising popularity of the CCF in check.[17] The CCF's policy of universal state ownership in the electrical

15. This table summarizes data provided in the Rural Electrification Committee's *Progress Report*, 1944, pp. 28–31.
16. BCARS, GR 1222, Premiers' Papers, Box 171, File 8, "Press Release, Premiers' Office," June 18, 1943.
17. In the 1941 election, the CCF received the largest portion of the popular vote, but this did not translate into the largest number of seats: CCF: 33.6% (fourteen seats), Liberals: 32.94% (twenty-one seats), and the Conservatives: 30.91% (twelve seats). See Robin, *Pillars of Profit*, p. 51.

industry – one central plank of its state socialist program – served to push the formerly noninterventionist Liberal–Conservative Coalition toward some public–private mix, at the very least. This leftward shift in policy paralleled a variety of Coalition attempts to engage reform issues as their own.[18] Furthermore, the hinterland focus of rural electrification fit well with the Coalition government's interest in opening new northern and interior regions to resource development after the war.[19] With the expansion of resource trade during the war, the possibility of opening BC's northern resource frontier to international capital loomed large in public discussion. State-led rural electrification could be envisioned not as a great departure for the parties of the center and right, but as an accompaniment to such government activities as road building. Ideology did not have to change, only the assumptions about the proper tools to be used by the state in assisting private capital accumulation.

The committee of civil servants created in 1943 to consider rural electrification was composed of W. A. Carrothers (chair), J. C. Macdonald, and Ernest Davis, who grasped this problem from the start and attempted to instruct politicians and the public in the possibilities and limitations of BC's existing electrical infrastructure. The first two members of the committee held a firm knowledge of the electrical scene through their posts in the provincial Public Utilities Commission (PUC) (Carrothers since its founding in 1938, MacDonald since 1939).[20] Before the war, they had overseen a major study of the province's private utilities with a view to mapping electrical rates and capital investments in order to determine the fairness of consumer costs. Far from being radical interventionists, members of the PUC, for the most part, shored up the claims of the dominant utility, the BCER, and judged its rate of profit to be fair.[21] They came at the issues of public power and rural electrification

18. Ibid., p. 78.
19. Wedley, "Laying the Golden Egg."
20. J. C. MacDonald had also been comptroller of water rights for the province from 1926 until 1939 when he joined the PUC: BCARS, GR 1006, BC Water Rights Branch, Box 1, File 1, "Board of Investigation and Water Branch Administrators, 1909–1965."
21. The study, commenced in 1939, finally reported to the public in 1943: BCARS, GR 1160, BC Public Utilities Commission, Engineering Department, Box 1, Public Utilities Commission, "Report to the Lieutenant Governor in Council on the Investigation into the Rates and Service of the British Columbia Electric Railway Company Limited and Associated and Subsidiary Companies," July 1943.

with a jaundiced eye, seeing in them the possibilities for great government expense and dubious outcomes. It is possible that Ernest Davis also had an impact on the committee's thinking, but the scant attention to water development in its reports would suggest otherwise. It is probable that the REC operated as an extension of the PUC and shared its assumptions.

The program recommended by the REC early in 1944 walked a fine line between state intervention and support for private utilities.[22] It focused on the organizational structure of the existing industry as the main obstacle to expansion. Unlike boosters such as Warren, the REC argued that state-led hydroelectric projects were not the answer, nor was the development of a province-wide grid. Electrical supply should grow modestly with market demand; it should build on the existing infrastructure. What the committee was recommending, in short, was a rationalization and integration of the existing medley of hinterland utilities. Economies of scale would provide a base for extension, rates could be dropped, and, with time, supply increased. Although state expropriation of the nonmetropolitan market might be one way to achieve this outcome, the report pointed out on several occasions that it was not the form of ownership, but the management style that determined the price structure of a given utility.[23] The report thus provided the coalition government with a technical strategy, but left open the question of how to obtain this outcome politically. Private utilities and public power advocates could both find reasons for cheer and concern in the REC's findings.

Private Power and the War

Private power interests followed the activities of the Post-War Rehabilitation Council and the REC with trepidation. BC Electric directors, in particular, understood the politically pleasing ambiguity of the term rural electrification. Perhaps it did not portend difficulties; perhaps it did.

22. *Progress Report of the Rural Electrification Committee as of January 4, 1944* (Victoria, 1944).
23. Ibid., pp. 16, 19.

Throughout the 1930s, BC Electric had sought a close match between the flattening demand of the electricity market and its own supply position. In 1931, the much-anticipated Bridge River project was shelved indefinitely when the Depression cut into Vancouver's industrial and domestic electrical market. Only in 1938 would extra capacity be added to the metropolitan Vancouver system with an upgrade at the Ruskin dam facility – accounting for a mere 47,000 hp.[24] Thus although capacity was added over the 1930s, the rate of growth was slower than in previous decades and the actual installation of additional horsepower was less.[25] Such a policy could be justified in uncertain times when demand was soft. But when war arrived in 1939, the company discovered the perils of forecasting on the basis of slow-growth assumptions.

Vancouver changed with the war, and so did its electrical market. Wartime needs brought forth an unprecedented level of shipbuilding, and airplane and armaments production, with all of the associated commodity processing.[26] As the city returned to full employment, the utility found streetcar ridership sharply increasing, placing a greater pressure on the daily load factor.[27] Although domestic demand was kept in check with dimout restrictions and BC Electric's own decision to halt the sales of electrical appliances in 1942, there was an overall growth in Vancouver's population, not to mention the additional draw of soldier encampments on the edge of the city.[28] Until 1942, the company could cope with these changes. It was only when the BPA requested a purchase of 90,000 hp in the spring of 1942 that BC Electric directors reassessed their position.[29] They could not meet the request, that much was

24. BCARS, GR 1289, BC Water Rights Branch, *Hydro-Electric Progress in Canada*, Department of Mines and Resources, 1937, p. 2.
25. R. C. Farrow provides aggregate figures of provincial installed capacity at: 65,000 hp in 1910, 310,000 hp in 1920, 630,000 hp in 1930, and 789,000 hp in 1941: Farrow, "The Search for Power," p. 89.
26. BCARS, GR 1289, BC Water Rights Branch, *Hydro-Electric Progress in Canada*, Department of Mines and Resources, 1942, p. 2.
27. Ewert, *The Story of the BC Electric Railway Company*, p. 225.
28. UBC, BCER Papers, Box 79, File 1583, "W. G. M." [President Murrin] to H. J. Symington, power controller, Department of Munitions and Supply, May 13, 1942 (copy). Murrin describes the impact of the war on the electrical business in this letter.
29. UBC, BCER Papers, Box 79, File 1583, diary of J. A. Brice, May 13, 1942, reports discussion of the Bonneville Power Administration request and the inability of the company to meet it.

obvious. But how many new needs would be placed on their system? How long could the company's existing supply meet the unprecedented rising demand?

BC Electric management believed that the most feasible expansion plan was the redevelopment of the Bridge River project, abandoned a decade earlier. A small system was already in operation, supplying a local mining venture, and some aspects of the earlier expansion work were complete. To add the facility's capacity to the metropolitan system, transmission lines, new generators, and an overall expansion plan would be required. This would be costly and require a host of scarce materials and labor, but it might be completed within two years. The problem was to obtain government priority, so that the envisioned supply of 500 miles of copper transmission wire, for example, might be forthcoming. For this, Ottawa's indulgence would be necessary.

To receive government priority for its construction needs, BCER had to turn to the Department of Munitions and Supply and make a case that the expansion was a wartime necessity.[30] During the war, a power controller in the department assessed the energy outlook across the country, imposed rationing methods where necessary, and approved or postponed hydroelectric construction projects on the basis of forecasted demand and a sense of national priorities.[31] BC Electric's appeal thus stressed rising wartime demand (up by over 14% in Vancouver in the first six months of the year) and underlined the risks to the power supply.[32] The company proposed as the solution a $6.5 million project to be sited at Bridge River and completed by September 1, 1945. Authorities at the Department of Munitions and Supply and the power branch of the War Production Board considered the request but turned it down because they judged that it would not be completed by the end of the war.[33] It could not therefore

30. UBC, BCER Papers, Box 79, File 1583, Copy of Application to the controller of construction, Ministry of Munitions and Supplies, nd; "W. G. M." [President Murrin] to H. J. Symington, May 13, 1942 (copy); "W. G. M." [President Murrin] to W. E. Uren, director general, Priorities Branch, Department of Munitions and Supply, August 20, 1942.
31. De Nevarre Kennedy, *History*, p. 181.
32. UBC, BCER Papers, Box 79, File 1583, "W. G. M." [President Murrin] to H. J. Symington, July 16, 1942 (copy).
33. UBC, BCER Papers, Box 79, File 1583, C. L. Rogers for W. E. Uren to W. G. Murrin, September 3, 1942.

be considered a wartime necessity and would receive no priority license.[34]

Embittered by this response, BC Electric management canceled the expansion plans and hoped that their predictions would not come true. A year later, in the spring of 1943, company managers attempted to devise a new course; an internal committee of engineers and managers discussed future war problems and the readiness of the firm for a return to a peacetime market. The wartime conditions, reasoned E. H. Adams, BC Electric vice president, would continue to eat into BC Electric's supply. Mobilization for the Pacific theater loomed as a possibility.[35] As to the postwar situation, a number of internal studies forecast a steady growth in industrial demand from the forestry-processing sector, among others, and a jump in domestic demand.[36] Adams also believed that the PUC would be looking to impose a new rate structure on the province with sharply lower domestic pricing. Experience elsewhere suggested that a drop in rates was usually accompanied by a rise in consumption.[37] Both the short- and the long-term outlooks therefore pointed to the need for additional capacity.

There was, of course, the further consideration of political forecasting. Public discussion of a state role in hydroelectric development and the activities of the REC had gained the attention of BC Electric managers. Additional capacity, they hoped, might also solve a public relations problem. "The development of a sizable block of power," explained BC Electric President Murrin to Montreal Director A. G. Nesbitt, "would put us in a much stronger position successfully to meet agitation for public ownership." Up until now, he stated, public reports of hydro development elsewhere were frequently accompanied by unfavorable comparisons with BC

34. UBC, BCER Papers, Box 79, File 1583, C. L. Rogers for W. E. Uren to W. G. Murrin, September 3, 1942; J. H. Gain, executive assistant to controller of construction to Bridge River Power Company, September 9, 1942.
35. UBC, BCER Papers, Box 79, File 1583, diary entry for A. E. Grauer, "Meeting of the Post-war Construction Committee held in board room," May 3, 1943.
36. UBC, BCER Papers, Box 79, File 1583, "Report on Post-war Activities: General Sales Department – Lower Vancouver Island," June 18, 1943; diary entry for A. E. Grauer, July 14, 1943, "Meeting of the Post-war Construction Committee held in the board room," July 14, 1943.
37. UBC, BCER Papers, Box 79, File 1583, diary entry for A. E. Grauer, "Meeting of Post-war Construction Committee held in the board room," May 3, 1943.

Electric. If a development program could proceed, the company would be able to protect itself from the accusation of "lack of initiative."[38] To expand defensively, however, Ottawa's assistance would be needed.

In the spring of 1943, BC Electric management attempted a new strategy. They bypassed the bureaucracy and headed straight for the minister of munitions and supply, C. D. Howe, to emphasize the seriousness of the problem and recommend radical solutions. Without government priority and financial assistance, they claimed, growth in the Vancouver market owing to an expansion in the Pacific theater of war would lead to serious supply and distribution problems.[39] Howe and his assistants reconsidered the request – which had in the intervening period doubled its budget – but turned it down once more.[40] They had consulted military staff and concluded that the talk of massive expansion in Vancouver because of the Pacific war was unlikely. Howe's power controller, Herbert Symington, offered assistance to the company in terms of depreciation on a single generation unit (not two, as had been asked for) as well as priority in obtaining supplies. But that was all. He did not explain what his department proposed to do if the threatened shortages arrived.[41]

This decision effectively killed BC Electric's immediate plans for expansion. In the summer of 1943, right in line with predictions, or even in advance of them, Vancouver suffered power shortages. An unusual stretch of dry weather had left BC Electric's reservoirs lower than normal, thus reducing their generating capacity. The company's usual strategy of augmenting the system with expensive thermal generation at the Burrard Inlet plant was impossible, given oil shortages. Instead, the company had to buy a block of power from a U.S. supplier, the Puget Sound Light Company, extending its reach not only beyond the basin, but also beyond Canada.

38. UBC, BCER Papers, Box 79, File 1583, "W. G. M." [President Murrin] to A. J. Nesbitt, June 1, 1943 (copy).
39. UBC, BCER Papers, Box 79, File 1583, "W. G. M." [President Murrin] to C. D. Howe, June 3, 1943 (copy).
40. UBC, BCER Papers, Box 79, File 1583, Adams to Murrin, June 16, 1943.
41. UBC, BCER Papers, Box 79, File 1583, Symington to Murrin, July 30, 1943. BCER argued with Symington about this decision and received a blunt rebuke: Murrin to Symington, August 11, 1943 (copy); Symington to Murrin, September 1, 1943.

Further, the company introduced new lighting restrictions in the city above and beyond those created for the entire dominion the year before.[42] All the while the Bridge River project remained idle. Engineers began surveys and consulted possible contracting firms, but no shovel was turned before the end of the war; new power from Bridge River would not be delivered to Vancouver until 1948. Instead, the city would hold on with the assistance of a power tie-in with the BPA starting in 1946. The BPA's appeal for assistance in 1942 had triggered BC Electric's quest for expanded capacity on favorable terms. Now, at war's end, the BPA bailed BC Electric out of its unenviable position. But was this good enough to hold off the agitation for state expropriation?

Public or Private Power?

Although the Coalition government supported the general concept of creating a public power commission of some variety after the war, the actual design and mandate of such an agency remained an open question. The Post-War Rehabilitation Council's interim report advised a cautious approach: Appoint a commission to study the possibilities of state intervention. To some extent, the REC fulfilled this goal, while nudging the government toward a more conservative approach than earlier proposed. The problem was that more ambitious plans had already created an expectant public audience.

From small beginnings, reflected in the briefs to the Post-War Rehabilitation Council in 1942, the public power issue developed momentum by 1944, attracting a coalition of interests calling for state intervention. Despite the fractious party politics of the day and the CCF's best efforts, the idea of public power resisted narrow party or ideological definition. While CCF politicians tried to monopolize the issue for the left, the Coalition government could respond that public power was a cherished ambition of free-enterprise

42. On the 1943 shortage, see de Navarre Kennedy, *History*, p. 185. On the tie-in with American sources, see BCARS, GR 1289, BC Water Rights Branch, *Hydro-Electric Progress in Canada* (1943), Department of Mines and Resources, p. 2.

government as well.[43] As a political concept public power was plastic: From the right it could be justified as a means to efficient service, from the left as a step to state socialism. Thus the initial public response to the Post-War Rehabilitation Council contained briefs from numerous boards of trade, chambers of commerce, and union locals as well as veterans', women's, and other service organizations. The political persuasions of these groups apparently ran the gamut. Despite the preponderance of hinterland over urban support in the council's findings, public power also resisted easy regional definition. Through the war, the largest dailies in Vancouver supported a broad program of public power. Urban-based labor groups, including the Vancouver Trades and Labour Council, also weighed in favorably on the issue, as did a host of municipal organizations.[44] As the question of the appropriate postwar course came to a head in 1944, municipal governments across the province entered into the debate in an attempt to transform state intervention from a narrow rural program to one that encompassed the province as a whole.

The municipal role reflected the breadth of support for public power across the province, but also portended difficulties in the implementation of a state-controlled system. Municipal involvement in this debate grew out of a long tradition of antagonism between municipal governments and monopoly utilities, particularly in the more urbanized regions of the province. In the summer of 1944, Mayor Cornett of Vancouver led municipal leaders in the formation of a broadly based civic public power coalition. A conference convened among cities and towns serviced by BC Electric produced a final resolution that called on the provincial government to expropriate the power system under an independent

---

43. Robin, *Pillars of Profit*, p. 73.
44. BCER CF, "Develop BC Industries, Take Over BC Electric, Establish New Industry is Demand at PP Meeting," *Pacific Advocate*, nd; "Labour Urges Public Power," *Pacific Advocate*, November 25, 1944; A number of different labor groups communicated with the premier, calling for a public utility: BCARS, GR 1222, Premiers' Papers, Box 172, File 1, John Turner, executive secretary of Vancouver Labour Council to premier, November 27, 1944; A. E. Papke, secretary of the International Woodworkers of America, District 1, Local 424 to premier, August 18, 1945. This letter states the local's support for a provincial utility.

hydroelectric commission.[45] Of the twenty-nine municipal councils represented, eighteen supported the resolution unconditionally, while eleven others supported it with minor qualifications.[46] A subsequent meeting held by the Okanagan Valley Municipal Association considered a similar proposal, with six of nine municipalities in favor, one undecided, and two against because of preexisting municipal ownership.[47] This broad-based municipal support – representing metropolitan and hinterland municipal councils – suggested that municipalities were ready to hold the provincial government to a broader meaning of public power.

Before municipalities and the province could agree on shared responsibilities, however, the grounds of discussion changed. In the fall of 1944, BC Electric announced that it would invest $50 million in a major overhaul of its utilities at the end of the war. This would allow for the development of delayed projects, like the Bridge River facility, and provide infrastructure for rural electrification in areas around BC Electric's urban markets.[48] With the province and the municipalities unable to agree on the division of powers under state control, BC Electric's defensive move put the public power campaign on hold. Dailies sympathetic to state expropriation chided BC Electric for its late conversion to a progressive agenda and claimed that the move was a cynical attempt to extract more money from the province in an eventual takeover purchase.[49]

45. The resolution put to the municipal councils for a vote stated "That the Council of. . . . a Municipality served by the BC Electric system, (on the assumption that any municipal revenues now accruing in respect of the Company shall not be impaired) goes on record as being in favour of the Province taking over and operating through an independent Commission the whole of the affiliated Companies' undertaking without participation by the municipalities."
46. BCARS, GR 1222, Premiers' Papers, Box 171, File 8, Mayor J. W. Cornett, Vancouver, to Premier Hart, July 25, 1944.
47. BCARS, GR 1222, Premiers' Papers, Box 171, File 8. The balance of municipal support in the Okanagan previously listed is a summary of separate letters sent to the premier by J. W. Wright, honorary treasurer, Okanagan Valley Municipal Association, August 30, 1944.
48. BCER CF, Copy of President Murrin's press release, September 26, 1944; "BE Electric Has Ambitious Plans," *Financial Times*, October 20, 1944; "$50,000,000 Program for BC Electric," *Financial Counsel*, October 12, 1944; "BCER's Expansion Plans," *Province*, October 2, 1944; "BCER Planning an Outlay of $50,000 Improving Plants," *Vancouver Sun*, September 30, 1944.
49. BCER CF, "Belated Repentance by BCER," *Vancouver Sun*, October 2, 1944; "BCER's Expansion Plans," *Province*, October 2, 1944.

Others suggested that a state program should move forward now only if it could be proven to be superior to the enhanced private system.[50] Instead of introducing expropriation legislation, as had been intimated around this time, the provincial government and the municipalities agreed to establish a commission to assess BC Electric and recommend a public power administrative structure.[51] What such a commission would discover beyond what the PUC already knew was unclear. Critics of the move dubbed it the "board of delay."[52]

Rather than wait for this new commission's findings, the province bypassed the process and pursued its own modified agenda. With no resolution to the BC Electric expropriation issue in sight and continuing municipal intransigence, the provincial government introduced legislation to establish the BC Power Commission (BCPC) in the late spring of 1945. The BCPC received a mandate to extend and expand electrical services in the nonmetropolitan regions of the province.[53] It had no authority to supply power in areas already serviced by BC Electric or the second largest private utility, West Kootenay Power. This left BC Electric with its urban empire in place and handed the public utility the job of extending services in less profitable markets. Months before a provincial election, the promise to put a "light bulb in every barn" proved to be "excellent electoral fare" in the hinterland areas of the province.[54] In the urban Southwest, BC Electric threw its considerable corporate weight and advertising budget behind the government's plan.[55]

In the aftermath of the coalition government's reelection in October 1945, provincial and municipal politicians continued to speak

50. BCER CF, "BCER's Electric Vision," *Victoria Daily Times*, October 2, 1944; "Utilities Expansion," *Vancouver News Herald*, October 2, 1944; "Utility Responsibility," *Vancouver News Herald*, November 9, 1944; "Looking Forward," *Victoria Daily Colonist*, October 3, 1944.
51. BCER CF, "Hart's Plan Finding Favor," *Province*, December 20, 1944; "Mayor Mott Firm for BCER Deal," *Vancouver Sun*, December 13, 1944; "Local Board to Operate Utilities," *Vancouver Sun*, December 11, 1944.
52. BCER CF, "A Board of Delay," *Province*, December 9, 1944; "Why Take Over BCER?" *Vancouver Sun*, December 5, 1944.
53. BCER CF, "Vast Power Distribution Schemes Announced by Province, and BCER," *Province*, February 5, 1945; "Electricity for Most BC Farms," *Vancouver Sun*, February 5, 1945.
54. Robin, *Rush for Spoils*, p. 80.
55. Robin, *Rush for Spoils*, p. 84.

in favor of public power while trying to insulate their governments from financial responsibility. The so-called Board of Delay reported in August 1945 and found substantially in favor of the provincial position.[56] Armed with this expert advice, which had been jointly called on by both levels of government, Premier Hart cornered municipal politicians. If they were not willing to ante up with the necessary financing for the expropriation of municipal utilities and their operation, he argued, then municipal referenda should be held on the issue to let the electorate decide the matter once and for all.[57] Mayor Cornett of Vancouver replied to the premier that there was little point in calling the question so long as municipalities were being asked to assume costs they could not afford.[58] The municipalities could not move on this point without risking their financial stability. With no agreement on financing and shared responsibilities, the public power issue collapsed in the province's urban regions.

The conclusion of the public power issue related to the restatement of the coalition government's goals in the aftermath of the 1945 election. Whereas before the election, the province feared the rising popularity of the CCF and sought to present public power as a central plank of a reform agenda, after having been returned with a majority, the attempt to fill the center in the political spectrum faded. The creation of the BCPC fulfilled its purpose: firming up nonmetropolitan support while laying the groundwork for hinterland industrial development. The threat of public expropriation in the cities, on the other hand, also delivered desirable outcomes: The BCER was now committed to a reinvestment agenda and promised massive expansion. The coalition had gone to the edge and reaped political rewards. Now reelected, Premier Hart extracted his government from the public power issue while portraying the municipalities as the unwilling partners.

56. BCARS, GR 1222, Premiers' Papers, Box 171, File 7, "Report on Proposed Acquisition of Properties of British Columbia Power Corporation, Ltd," August 28, 1945.
57. BCARS, GR 1222, Premiers' Papers, Box 172, File 7, Hart to Cornett, November 6, 1945; Hart telegram to Cornett, November 7, 1945; Cornett telegram to Hart, November 7, 1945.
58. BCARS, GR 1222, Premiers' Papers, Box 172, File 7, Cornett to Hart, November 8, 1945.

Public and Private Expansion

The resolution of the wartime public–private power fight unleashed an unprecedented hydroelectric building program in BC. BC Electric built a transmission line connecting metropolitan Vancouver to the BPA's grid in 1946 and began construction on the much-postponed Bridge River facility in the same year. The newly formed BCPC proceeded in 1945 to purchase a host of small electrical systems on northern Vancouver Island, the Okanagan, and the interior. As well as improving plant and transmission infrastructures, in 1947 the BCPC launched new hydroelectric projects on northern Vancouver Island and in the Okanagan.

BC Electric's inability to coax Ottawa to extend priority and later financial incentives to expand its metropolitan supply system during the war placed the utility in dire straits by 1945. Demand continued to climb even as wartime production contracted. The provincial PUC went so far as to insist that the company delay no longer in expanding its supply system by constructing the Bridge River facility.[59] BC Electric, understandably, had delayed expansion plans in part to avoid the high costs of materials and labor in 1945, but also no doubt because of the threat of public expropriation. Its political problems seemingly averted with the formation of the BCPC in the summer of 1945, BC Electric began to deliver on its promise of a $50 million postwar expansion program.

BC Electric's program focused on the development of the Bridge River facility. Located near the Fraser Canyon, the project diverted water from the Bridge River – a cold, turbid, glacier-fed stream – through Mission Mountain and down a steep slope into Seton Lake. Using a relatively small volume of water, the diversion took advantage of a sheer drop between neighboring watersheds and operated at a high head. First surveyed in 1912, the project had reached a preliminary stage of completion in 1934 to produce power for the locality. Expansion in 1945 entailed building new storage and diversion dams in the upper Bridge River valley, installing new generators at the powerhouse on Seton Lake and connecting the

59. BCARS, GR 1222, Premiers' Papers, Box 172, File 7, Copy of BC Public Utilities Commission Order, June 6, 1945; File 2, W. A. Carrothers, PUC, to premier, May 16, 1945.

project to the urban market through transmission lines. Estimates in 1947 suggested that the project would deliver 186,000 hp on final completion.[60] When construction contracts were let in 1946, BC Electric hoped that the new power supply would reach Vancouver within two years. Although the building phase of the Bridge River project proceeded apace, the question remained as to whether it could come on line before Vancouver's metropolitan system faced a supply crisis.

BC Electric engineers judged not. Early in 1946, the company obtained Canadian and American authorization to construct a transmission line from its Vancouver system to the border to create an interconnection with the BPA's grid.[61] The new connection provided the firm with enough flexibility to complete the Bridge River project without resorting to drastic rationing methods or risking brownouts. Although the immediate problem was supply, BC Electric also envisioned using this link to sell excess power from the completed Bridge River facility to the BPA in the late 1940s. The transmission line was the first step in the creation of a transnational supply-balancing system between the two jurisdictions.

The Bridge River project also had important local effects. The development work at Bridge River, completed in 1934, had caused resentment among local native bands whose reserve lands were transgressed and crossed by new infrastructure; the expansion work did not rectify the problem.[62] New dams flooded reservoirs that displaced several small farms and removed former native hunting and gathering sites. Because development occurred in the upper Bridge River Basin where low water temperatures vitiated significant spawning habitat, dam development did not appear at first to harm local conditions for fish. After several years, however, salmon runs in the point of water displacement, Seton Lake, experienced declines. Owing to the transfer of large volumes of cold, turbid

---

60. British Columbia, *Lands, Surveys and Water Rights Branch Annual Report* (1947), p. 130. The initial installation to be completed, however, would be 62,000 hp. The Department of Mines and Resources, on the other hand, provided an "ultimate capacity" figure of 600,000 hp: BCARS, GR 1289, BC Water Rights Branch, *Hydro-Electric Progress in Canada* (1945), Department of Mines and Resources, p. 2.
61. The project was announced late in 1945: BCER CF, "BC Enters Power Pool Next Year," *Vancouver News Herald*, November 8, 1945.
62. Drake-Terry, *The Same As Yesterday*, Epilogue.

water from the Bridge River Basin to Seton Lake, limnological changes, in particular depleted phytoplankton and zooplankton, diminished the capacity of the lake as a salmon-rearing habitat.[63] The lake had once been a significant site for salmon spawners; it became less so. The native catch of returning spawners on the Seton River was reduced accordingly.

While BC Electric focused on developing physical infrastructure, the BCPC sought to consolidate an organizational framework for hinterland development. The commission aimed to improve supply and availability and increase the number of electrical districts and consumers. Under the leadership of Samuel R. Weston, an engineer formerly with the PUC and the REC, the BCPC followed a relatively cautious approach, initially expropriating three small private utilities in five months and creating two electrical districts, one on northern Vancouver Island, the other in the Okanagan. From these two points, the expropriation of other small systems and utilities continued in the following years, developing by 1949 into two major regional organizations: one coastal, the other weighted in the interior. The BCPC even succeeded in shearing off sections of the BC Electric empire by expropriating the company's Kamloops utility. BC Electric did not protest; it was a minor concession in order to be left alone in the major markets.[64] The development of the system occurred in this piecemeal fashion rather than in one fell swoop, in part because the provincial funding formula covered only initial capital costs. The BCPC had to generate its own administrative and operational costs from revenues and so it could not afford rapid growth. Within a few years, the integration of smaller systems produced twelve separate electrical districts, some with sufficient supply to offer promotional rates that helped to spur the expansion of domestic consumption – "one of the main objectives of the commission," as the 1949 Annual Report put it.[65]

63. G. H. Geen and F. J. Andrew, "Limnological Changes in Seton Lake Resulting from Hydroelectric Diversions."
64. BCER CF, "Vast Expansion of BC Power System Seen," *Province*, October 24, 1946; "BCER Won't Fight to Keep Power Plants," *Vancouver Sun*, October 23, 1946; "BCE Properties to Cost BC $1,400,000," *Vancouver Sun*, October 22, 1946.
65. The early expansion program of the BCPC and the limits on growth are well covered in the Commission's Annual Reports from 1946 to 1949. The quotation is from p. 11 of the 1949 Annual Report.

Early in its development the BCPC began to investigate hydro-electric possibilities. Although on northern Vancouver Island and in parts of the interior the BCPC ran hydro facilities, many parts of the interior depended on costly thermal generation. Building on its core markets, the commission began its development phase by building hydro projects on Campbell River on Vancouver Island and at Whatashan on the Arrow Lakes in the upper Columbia Basin to supply northern Vancouver Island and the Okanagan, re-spectively.[66] After the creation of these regional growth spurs, the BCPC found that the combination of increased supplies and re-duced rates propelled domestic and commercial consumption. As the Campbell River development opened December 15, 1947, with John Hart flicking the proverbial switch, plans were already under-way for its expansion.[67]

The River as "Destroyer"

As if soaring consumer demand and promotional rates were not enough to drive the postwar dam-building program, a new fac-tor arose in the spring and summer of 1948. The Fraser River flooded. Starting on the holiday weekend of May 24, the Fraser River climbed its banks, pushed back feeble containing walls and dikes and charged forth across the landscape, flooding agricul-tural lands, destroying homes and severing all of Vancouver's land links with the rest of Canada.[68] Although since 1858 the river had flooded on average once every four years, not since 1894 had the river risen on such a scale.[69] The combination of a heavy snow-pack

66. BCER CF, "Island Power Project Ready by Next Spring," *Vancouver News Herald*, July 15, 1946; "Famed Falls Survive Big Island Dam," *Vancouver Sun*, December 16, 1947; "More Power to BC," *Vancouver Sun*, December 17, 1947; "Big New Power Project Announced for Interior," *Victoria Colonist*, December 16, 1947.
67. The dam opening is described in BCER CF, "Famed Falls Survive Big Island Dam," *Vancouver Sun*, December 16, 1947. On the plans for expansion capacity, see "Power Scheme Doubled," *Province*, March 8, 1947; "The Public Demanded It," *Victoria Daily Times*, February 19, 1947. The building program of the BC Power Commission in the late 1940s is described in John B. Shaw, "The BC Power Commission."
68. This account of the flood is based on NAC, RG 89, Box 654, File 2258(B), C. E. Webb, district chief engineer, Department of Mines and Resources, "Flood of 1948 in British Columbia," Vancouver, BC, March 15, 1949. For a set of journalistic portraits of the flood that emphasize the cohesiveness of communities in the face of the disaster, see *Nature's Fury*, available at BCARS.
69. Sewell, "Changing Approaches," p. 105.

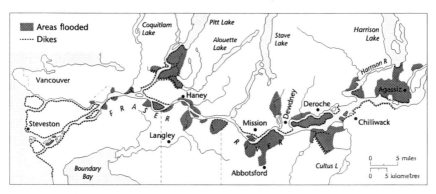

Map 5. Flooding in the Fraser Valley, 1948.
(*Source:* Fraser River Board, *Preliminary Report on Flood Control.*)

in the coastal and interior ranges in the winter of 1947–8, sustained low temperatures in the months of March, April, and early May – when the river usually began to drain the mountains of their icy load – and a sudden rise in temperatures in late May created the conditions for an enormous outpouring. The first sign of trouble was the overflow of Bonaparte Creek near Cache Creek. Then, on May 25, meter readings at Mission showed the river at a height of 19.5 ft, a mere half-foot below the known level of flood danger. The following day dikes began to break in the Fraser Valley (see Map 5). At Hope the river disgorged 401,000 ft³/s. Before the flood subsided, the water would climb above 26 ft at Mission and an additional 135,000 ft³ of water would be added to the earlier balance of 401,000 passing every second at Hope. In 1894, when the river rose to an even higher level, the flood passed quickly. In the spring and summer of 1948, by contrast, the river stayed at flood levels for thirty-three days, a flood duration not matched in the historical record, or since. Finally, by June 26, the river receded below the 20-ft mark at Mission, and the restoration of the valley could begin.

The inundation produced a stream of damage statistics: 70,000 acres flooded, 2300 homes damaged or destroyed, 16,000 persons evacuated, $20 million in damages. The brunt of the impact was felt in the Fraser Valley from Aggasiz to the delta. One-tenth of the valley's land base, or 50,000 acres, went under water. A quarter of the diked land in the valley was inundated. First roads, then the CPR line, and later the CNR line were battered and severed. Airlines began additional shuttles to Vancouver, carrying supplies. Some 30,000 civilians engaged in relief work, and 3000 military personnel led a centralized flood-control program. One small pile-driving firm repaired fifty bridges on the Fraser before the year was out.[70] When the press hailed the event as one of the most remarkable environmental catastrophes in the history of the province, they for once did not engage in overstatement.[71]

Floods, as students of natural hazards argue, have a remarkable power to focus public discussion on significant changes in water management.[72] As the flood coursed across the valley floor, metropolitan dailies began the process of airing possible solutions to the devastation.[73] Editorials and opinion articles unanimously suggested the need for an agency of experts to oversee integrated planning in the river basin. Here again, American precedents provided a conceptual vocabulary: Many articles leaned toward an authority like the TVA that might bring the Fraser's different interests together.[74] Just as on the swollen Columbia in 1948, the excuse of a flood was quickly deployed to promote dam construction.[75]

70. *Fraser River Pile Driving*, p. 24.
71. BCER CF, "Governments Take Action in BC Flood Battle," *Province*, May 28, 1948.
72. However, as Derrick Sewell, a student of White's, would add, once crises pass, their role as a motivating force for action passes too: Sewell, *Water Management*. For a brief statement of White's approach, see White, "Comparative Analysis."
73. Alongside this long-range-planning discussion, the press spent much time praising the citizenry for pulling together in unexpected and pleasing ways: Mennonite farmers helping Army personnel sandbag a broken dike, for example, were images that gained journalistic attention. For a selection of this flood journalism, see the small compilation produced after the flood from a sampling of newspapers: *Nature's Fury*, May–June, 1948.
74. BCER CF "No More Floods," *Vancouver Sun*, June 1, 1948; "From an Angle on the Square," *Province*, June 12, 1948; letter to the editor from B. Whitten, "The Flood Question," *Vancouver Sun*, October 2, 1948; "IWA Urges Big River Program," *Province*, June 22, 1948; letter to the editor from A. Cheverton, "Food Rehabilitation," *Vancouver Sun*, June 19, 1948; "Flood Control and Power not Attuned," *Vancouver Sun*, June 17, 1948; "Lack of Foresight Blamed for Flood," *Vancouver News Herald*, June 16, 1948; "To Harness the Rivers," *Victoria Daily Times*, June 14, 1948; "Rehabilitation of Fraser Homeless Pledged by Johnson – Dams Urged," *Victoria Colonist*, June 13, 1948; letter to the editor from Donald Bruse, "Technocracy and Floods," *Vancouver Sun*, June 2, 1948; letter to the editor from 'Waterman,' "Electric Power and Flood Control Could Well Be Combined on the Fraser," *Vancouver Sun*, May 27, 1948; "Flood Control Means Power for BC," *Vancouver Sun*, May 21, 1948; "Are Dykes Without Dams Sufficient?" *Vancouver Sun*, September 25, 1948.
75. White, *The Organic Machine*, p. 75.

In the midst of the crisis, the *Province* provided a reflective history of the Fraser that cast the river in the role of province-builder-turned-province-wrecker[76]:

> It is not going too far to say that the Fraser was responsible for making British Columbia and that it has been a major factor in the province's development. But today it is a destroyer instead of a creator. Women and children are fleeing before its onset and men are toiling with bulldozers and trucks and sandbags in a vain effort to stay the spread of its waters. One doesn't have to be an engineer to know that the place to control the Fraser is not on its lower reaches and that the filling and piling of sandbags is a desperate and very temporary measure. The Fraser is a system rather than a river. Some of its tributaries are great rivers in themselves. Its lakes and creeks run into the thousands. The place to control the Fraser is up above, where the water comes from. The way to control it is to control the tributaries. For this we need a Fraser River Authority – a board with authority enough and resources enough to take the river in hand, control its transports, coordinate its industries and make it work consistently for the benefit of the province as a whole.... The authority's task will be to make the river work and keep it from destroying.

The sentiments expressed in this editorial, penned at the height of the flood, represented well the mood of the province and its politicians. On a number of occasions, Herbert Anscomb, Conservative leader and minister of finance in the Coalition government, informed the press that he believed a series of dams in the upper basin would be necessary to stop similar disasters in the future.[77] As the province and the federal government deliberated over how to divide the burden of the cleanup and recovery, politicians from both levels of government agreed to establish a federal–provincial board to study future needs in water management in the Fraser Basin. The Dominion-Provincial Board, Fraser Basin, created in the fall of 1948, brought together ten senior civil servants from a number of federal and provincial departments, hired a full-time engineering staff, and set to work on a five year mandate to investigate how best to prevent another flood. This new institutional authority

76. BCER CF, "Let's Make the Fraser Work, Not Destroy," *Province*, May 31, 1948.
77. BCER CF, "Anscomb Appeals for Power Plan at Fraser Headwaters," *Vancouver Sun*, June 22, 1948; "More Than Dykes Needed," *Vancouver News Herald*, June 15, 1948.

held only ad hoc powers (unlike the TVA), but would nevertheless play an important part in the fish vs. power controversies ahead.[78]

Conclusion

"To cut a long story short – we missed the boat before the war, we have missed it during the war, and unless we take steps soon, we may miss it after the war."[79] Harry Warren's comment referred specifically to government waterpower surveying but it might have extended to the whole realm of water-development activities that he helped to inspire during the war. For this geologist and BC enthusiast, the rivers of the province represented just so much pent-up energy. From another perspective, the pent-up energy was more properly a social phenomenon of which Harry Warren was a conspicuous element.

The political economy of power changed during the war and after. In 1939, a dominant utility in urban regions and a scattering of over sixty small utilities in hinterland areas served the province. Electrical rates were relatively high, particularly in smaller centers; rural electrification was marginal. The state played only a limited role in the business: a PUC established in 1938 assessed the fairness of rates, but erred on the side of corporate interests. Despite BC's abundant waterpowers, no supply-driven building program appeared in the 1930s, despite the examples of New Deal projects on the Columbia. BC Electric followed instead a strategy that closely matched supply with demand. This approach avoided risky expenditures, but came perilously close in wartime to forcing the hand of the provincial government to expropriate the firm in order to expand the electrical supply.

Wartime unleashed new demands as well as expectations. In urban areas of the province, population growth and the expansion of industry consumed the marginal supply of BC Electric and forced it to examine immediate expansion options. Across the province, in

78. BCER CF, "Board Maps Plan to Study Fraser," *Vancouver Sun*, November 8, 1948; "Dominion-Provincial Board Set Up to Study Flood Control on Fraser," *Victoria Colonist*, July 3, 1948.
79. BCARS, GR 1222, Premiers' Papers, Box 165, File 11, Harry V. Warren, "Some Thoughts on Post-War Construction," nd.

hinterland areas and in the urban centers, disparate groups joined in calling for some form of state intervention in the electrical business. To stem the rising popularity of the left and reinforce hinterland support, the provincial government launched attempts to investigate rural electrification while leaving open the option of a broader policy of state expropriation. Although the provincial government cited the intransigence of municipalities as the reason it shifted its policy away from an ambitious program of province-wide expropriation, the greater factor was probably the rearguard action of BC Electric and its promise to expand electrical development in the province. In place of an integrated state system, British Columbians received a renewed monopoly of private power in the cities and a public utility in the hinterland regions. From 1945 to the end of the decade, a host of building projects studded the rivers of the province with new dams, power plants, and pipelines. Across watersheds and mountain ranges new power lines transmitted electricity to cities, towns, and, increasingly, hinterland areas beyond.

Remarkably, the sudden burst of dam building after the war did not impose a major burden on the river habitats of salmon. The Bridge River facility had effects on Seton Lake runs, but on a much smaller scale than if main-stem development had occurred. The BPA tie-in displaced environmental impacts on the power-generation region: the Columbia River. Although the John Hart dam was located on a famous salmon stream, the Campbell River, its initial impact appeared negligible. As in the period of hydroelectric dam development before the war, these results were products of fortuitous circumstance rather than conscientious environmental planning.

Yet, if the immediate impact of these developments did not disturb salmon, they helped to lay the conditions for future problems. The development of new power sources after the war provided the supply basis for a major expansion in electrical consumption in BC. Before the war, the high rates and poor service of the electrical business had helped to keep the electrical market underdeveloped and, by implication, less damaging to salmon habitat. After the war, new power sources allowed utilities to offer promotional rates and absorb new consumers, both domestic and industrial. In the first five years of its operation, the BCPC, for example, oversaw a threefold increase in electrical consumption in areas under its jurisdiction. This rate of growth was unprecedented in the province. New dams laid the ground for more.

The public power fight and the devastation of the 1948 flood also created new institutions that would weigh heavily in fish vs. power fights in the future. The BCPC, a small provincial utility by the late 1940s, had grand ambitions. Yet, because its market was relatively small, it proceeded according to a regional supply policy to avoid the overhead costs of a major grid network. This established a structural condition disadvantageous to river protection: Each region would need its own power supply. Rather than favoring large developments capable of major supply possibilities, the institutional design of the BCPC set it on course to develop small sites near points of consumption. The prospect of more dams meant increased opportunities for affecting salmon habitat. The Fraser Basin Board posed a different sort of problem. Its mandate concerned flood control, and a major point of study would be dam development. Although the decentralized character of the board between different government departments lessened its focus both in design and purpose, it appeared well positioned in 1948 to carry out a major flood-control program on the Fraser. Flood-control dams would require main-stem developments and upper-basin sites. This would place barriers directly in the course of migrating salmon.

By the end of the 1940s, the dam-building era in BC was in full stride. Projects were coming to completion; others were under investigation. It was difficult to imagine, if one were standing at the Bridge River powerhouse in 1948 with the political elite greeting the new day of power, that this released pent-up energy could be stopped.

5

The Power of Aluminum

From the late 1930s to the late 1940s, the contexts driving hydroelectric development in BC changed markedly. In the late 1930s,
small systems had existed in resource towns associated with primary industry. In the cities, relatively modest utilities had faced
the challenges of the Depression by reining in system expansion
and consolidating existing markets. A decade later, the tumult of
a world war had transformed the provincial political economy.
BC had experienced an intensification of resource and industrial
development, the population had grown, and new connections
tied the province into a wider continental economy. As part of
these changes, the provincial government had entered the water-
development and rural electrification business, private utilities had
forged ahead with building plans, and politicians and pundits had
begun to look to the northern interior to expand the provincial
resource economy.

The politics of this resource development agenda were both
local and international in emphasis. Small communities desired
electrical power for domestic needs and industrial development;
surveyors scrutinized small rivers to meet the needs of outlying
settlements; electricity linked sections of the province and tied together metropolis and hinterland in new ways. Yet all of these local
concerns were reflective of British Columbians' desire to catch up
with other regions, to modernize and to mimic power development
elsewhere. Local development was tied to the drive to expand the
province's reach in the international commodity markets. Hydroelectricity was increasingly seen as a magnet to draw international
capital investment. Overriding concerns of continental security in

an emerging Cold War environment conditioned the reception of development projects and shaped their meanings. Newspapers reported the progress of hydro dams built by the "Reds," just as editorialists called for an equivalent building program in Canada. Local and international concerns were thus linked; they implied each other.

In the immediate postwar years, small-town ambition, growing urban demand, and provincial planning had shaped the power-development agenda. By the late 1940s, a new politics of development had emerged with stronger ties to the international scene. Multinationals examined BC rivers for cheap waterpower possibilities. International military pressures recast BC's position in continental defense and as a supplier of resources. The Cold War heightened anxieties and provided a new language of debate. How would a formerly peripheral region of North America experience the influx of postwar development capital? How would the interests of different and conflicting resource users be weighed and managed? How would the politics of development join local and international concerns and transform them both?

Prospecting for Waterpower

From the air, Chilko Lake and the Nechako River must have seemed wild. Distant from the province's urban centers, well beyond the reach of highways, the lake sat against an impressive surround of mountains and meadows, and, farther to the north, the river charged through a dense northern forest of pine and poplar.

The calculating eyes taking in these views from a helicopter belonged to executives of the Aluminum Company of Canada (Alcan), touring water-development sites in the fall of 1948. Company President R. E. Powell joined provincial Minister of Lands and Forests E. T. Kenney to see the possibilities and publicize his company's interest. On the ground, Alcan engineering and survey crews had already begun to map the territories and consider possible development schemes. At Chilko Lake, provincial Water Comptroller Major Richard Farrow led Bill Huber of the International Engineering Company through the area. Farrow had surveyed Chilko Lake before the war and reported on its potential to the Royal Geographical Society in London in 1944; he now saw the possibility that earlier resource promotions might at last be realized.[1]

Chilko Lake and the Nechako River had long stood as the best prospects for water development in BC. In the 1920s and 1930s, the areas had been catalogued by the provincial government and advertised to would-be developers with glossy photographs emphasizing stirring white water. What made these places stand out were not the singular natures of their waterpowers – there were many other sites in BC that offered rushing rivers and storage in large lakes. Nor were the locations particularly advantageous. Chilko Lake sat northwest of the Fraser Canyon in rugged country, which until recently had seemed prohibitively distant from the provincial heartland. The Nechako River was even worse in this respect. It was located west of Prince George in the upper reaches of the Fraser Basin, surrounded in part by a provincial park. What made these sites different and potentially lucrative was that they lay near or parallel to other bodies of water flowing in the opposite direction toward the coast. If Chilko Lake or the Nechako River could be diverted west, piped to the coast, and delivered to powerhouses, they could provide large blocks of power at accessible coastal locations. Water from Chilko Lake could enter the Homathko River and produce power near Bute Inlet. A pipeline carrying the reversed flows of the Nechako could also reach a coastal location, within affordable transmission distance to a deep interior fjord, suitable for large ships. The sites nevertheless were still distant from urban centers. They would require industrial developers whose key locational concerns would be electrical supply, ocean accessibility, and connections with resource hinterlands, and who were prepared to make substantial initial investments in fixed capital. Before World War II there had been none. By the late 1940s, with an expansion in the postwar economy gaining head and new defense concerns reshaping the political economy of northwestern North America, the situation was about to change.

During World War II, the aluminum industry had experienced a period of considerable growth and expansion. Although classed as

1. Farrow, "The Search for Power"; Kendrick, *People of the Snow*, pp. 81–117.

a precious metal at the turn of the century, by the late 1930s aluminum was becoming important as a lightweight, durable product, well suited for airplane construction and a host of other military and civilian uses. Beginning in 1940, new military demands for aluminum in Britain and the United States created a serious supply crisis. In the past the dominant U.S. Alcoa Corporation had commanded the market, but it was no longer able to keep pace. In 1941, in an attempt to generate new supplies, as well as to break Alcoa's virtual monopoly control, the U.S. federal government intervened in the aluminum business, establishing its own plants as well as funding expansions in plant facilities of non-Alcoa interests. A number of smelters were established to consume the cheap power generated by federal dams on the Columbia River. By 1943 federal aluminum smelters consumed nearly two-thirds of the BPA's output.[2] Washington D.C.'s agenda reached beyond the United States. Alcan, a former subsidiary of Alcoa, but legally independent since 1928, received favorable American treatment under wartime pressures.[3] The United States federal government extended advance payments and loans to help fund a massive expansion at the company's Shipshaw facility in Quebec and signed contracts that became the envy of American producers at the end of the war.[4] Other allied nations followed suit. The Canadian government also deferred taxes on the company to assist rapid expansion. Between 1940 and 1943, Alcan increased its ingot-producing capacity fivefold; its aluminum plant at Arvida became the largest in the world. Measured in terms of sales and assets, Alcan experienced massive growth from 1937 to 1944: gross sales grew from $49 million to $259 million; assets jumped from $98 million to $523 million.[5]

2. The details of Alcan's dealings with the U.S. Federal Government and the controversy the deal created after the war is well covered in a chapter on "New Basic Industries for the West: Aluminum," in Nash, *World War II and the West*, especially pp. 103–5. For the statement of the portion of megawatts consumed by aluminum smelting, see White, *The Organic Machine*, pp. 72–3.
3. Alcan's relationship to Alcoa was complex and the subject of a number of different antitrust hearings in the United States that culminated with a 1950 decision that forced major bondholders in Alcoa to divest their control in Alcan. For a consideration of the relationship between the two firms, see Barham, "Strategic Capacity Investments."
4. Nash, *World War II and the West*, pp. 91–121.
5. Litvak and Maule, "Alcan Aluminum Ltd."

In the 1940s, the aluminum business operated according to a geographically dispersed production model driven by the costs of transportation and energy use.[6] The primary ore of aluminum is bauxite, a reddish-brown clay found in tropical and subtropical regions beneath layers of topsoil. Although in the 1940s most major sources of bauxite processed in North America originated in the Caribbean and South America, only rudimentary processing took place in these regions; at some locations bauxite was converted to alumina (aluminum oxide) before transhipment. Because ocean-shipping costs of bauxite and alumina were relatively low, aluminum producers could reap significant savings by moving these materials to sites of cheap energy supply in industrialized nations for smelting close to market. Aluminum smelting consumed massive amounts of energy – 20,000 kWh/ton of aluminum by one 1950s estimate – or twice the amount of energy that an average BC household in 2000 would use in a year.[7] By locating near to cheap power sources, the major component cost of aluminum production could be reduced and the final product delivered efficiently to market in industrialized centers. At a few points in North America, the two key locational factors of this production model – ocean accessibility and cheap waterpower – coincided, as on the Columbia and in Quebec's Saguenay River region. Some sites in BC possessed the same potential, but, before the war, remained little known.

Wartime shifts in the aluminum business were noticed in BC. In 1940, as the crisis in wartime aluminum supply took hold, UBC geologist Harry Warren was busy at work in his lab attempting to determine the costs of aluminum smelting in BC as part of his contribution to military-oriented research. He reported his findings to Alcan officials, invited them to tour the province, and connected them with the provincial mining department.[8] At the same time, federal officials encouraged Alcan to investigate BC's waterpower

6. Bunker and Ciccantell, "Evolution of the World Aluminum Industry."
7. Aluminum Company of Canada, "Kitimat-Kemano," p. 7. BC household estimates may be obtained from the BC hydro website: www.bchydro.com.
8. UBC, Warren Family Papers, Box 4, File 7, Warren to Kennedy, November 28, 1940; Kennedy to Warren, December 13, 1940; Warren to Kennedy, February 26, 1941; Kennedy to Warren, March 10, 1941; Warren to Kennedy, May 16, 1941; Kennedy to Warren May 23, 1941.

possibilities and Ernest Davis, the BC water comptroller, hosted visits by Alcan officials. Alcan planned no projects initially. In 1943, with the full extent of military demand unknown and construction at Alcan's Quebec facilities proceeding apace, the company's only plans for BC were for surveys.[9]

Alcan's reconnaissance activities were one aspect of a broader reconsideration of the geopolitical position and resource potential of BC and northwestern North America that occurred during the war. In the North, the American army connected sections of Alberta, BC, and the Yukon Territory to Alaska with a new highway. The Canol pipeline was constructed to facilitate the development of northern oil. Airplane reconnaissance activities increased. And a joint board of bureaucrats and planners from Canada and the United States began a survey of resource development possibilities in the region. The context of total war diminished national rivalries and opened the area for cooperative development. Looking toward the future, planners saw that the region would supply a host of raw and semiprocessed materials to the postwar economic recovery effort and would be an important defense zone against a possible Soviet invasion.[10]

Canadian economic integration with the United States continued after World War II, perhaps most vigorously in the mineral sector. In the United States, strong postwar domestic demand coupled with Cold War military spending created a ready market for Canadian mineral resources. By 1950, Canadian exports to the United States of aluminum, copper, lead, nickel, zinc, uranium, and iron ore had all surpassed peak wartime export levels.[11] In the case of aluminum, for example, Alcan grew rapidly during the war to supply British and American demand and continued to develop its business in the postwar period by using the comparative

---

9. The role of the federal government in encouraging Alcan to look west and the company's inquiries in 1943 are recounted in a brief summary of the project: BC Crown Lands Registry, Department of Lands 'O' Files, File 0124854, Alcan General (1), G. E. Melrose, "Memorandum to Mr. P. E. Richards, Executive Asst. to the Premier, Re: Aluminum Company of Canada Development," November 14, 1949 (copy).
10. Evenden, "Harold Innis"; Grant, *Sovereignty or Security?* pp. 129–87; Stacey, *Canada,* pp. 360–3.
11. Thompson and Randall, *Canada and the United States,* p. 203.

advantages of cheaper Canadian waterpower and labor. By the end of the decade, Alcan matched Alcoa in aluminum output capacity and attained a strong position in the North American market, despite a high Canadian dollar and punitive aspects of U.S. trade policy.[12] By the late 1940s, the firm was eager to grow. Executives sought cheap waterpower and proximity to markets in a politically stable setting. This was what brought Alcan officials to BC in 1948 to stare down at Chilko Lake and the Nechako River and imagine potential transformations.

Provincial policy favored the entrance of Alcan. The Coalition government placed great importance on opening the interior and north to development in the late 1940s. Since the war, government bureaucrats had advanced plans to expand the interior and northern highway networks, politicians urged international investors to examine the provincial timber wealth, and the nascent provincial electrical commission (the BCPC) aimed to expand power supplies in the southern and central Interior.[13] The possibility of an investment by Alcan complemented the planned development drive into the Interior and north and proved irresistible to provincial politicians. The development, it was hoped, might act as a growth spur for other projects. Alcan officials told the press that a company town located near a smelter facility might support a population of 50,000, which would make it the third largest city in the province. Extra power produced from the project might support a new pulp and paper industry in the region.[14] There was also Canada's contribution to continental security to be considered. Whereas before World War II, the province was unprepared for military engagement, now, with looming threats in the Arctic and Pacific, BC had to be mindful of its geopolitical position and its obligations to the nation and the continent. A new aluminum development in the province would help to solidify BC's role in whatever military and industrial endeavors lay ahead.

12. Smith, *From Monopoly to Competition*, pp. 286–7.
13. Wedley, "Laying the Golden Egg."
14. BCER CF, "World's Largest Aluminum Plant Proposed by Alcan on BC Coast," *Victoria Times*, December 17, 1948; text accompanying photograph of Kitimat site, November 21, 1949; "Aluminum Industry to Be BC's Biggest," *Vancouver Sun*, March 28, 1950.

The provincial government's support for Alcan's potential investment suggested a wholehearted engagement with continental defense and economic integration. The government, however, was aggressively provincialist. Before the Alcan investigations, the U.S. Bureau of Reclamation had surveyed a project in Alaska, to be fed by waters from the Yukon River that would entail consequences for water and land use in both the Yukon and BC. The BC government displayed a passing interest and engaged with the Canadian government in considering various proposals for what became known as the Taiya project. For the project to proceed, it would require Canadian approval for diverting headwaters located in Canada for an American development on the Alaskan coast. With Alcan's entrance to BC in the offing, however, the provincial government delayed and began to obstruct the proposal at the intergovernmental level. Rather than risk supporting a potential rival to Alcan's development in BC, the provincial government sought to undermine it. Alaska Governor Ernest Gruening viewed Canada's objections as rank obstructionism and argued that the Taiya project would produce more power and could be defended with greater confidence than a project in BC.[15] Notwithstanding the integration of this region forced by the war, by the late 1940s the rivalries between jurisdictions in attracting large hinterland development projects had returned with vigor.

To remove any doubt that the provincial government would support Alcan's arrival in BC, provincial officials explained to Alcan the province's level of interest. Following Powell's flying tour in the summer of 1948, E. T. Kenney, BC's minister of lands and forests, invited the company to state its needs and wants. To facilitate the company's planning, Kenney pledged to place reserves on the possible sites of interest to Alcan in order to exclude the intervention of competitors. He also offered the lowest possible water rental fees allowed under the law and concluded with this remarkable concession[16]:

15. Naske, "The Taiya Project."
16. BC Crown Lands Registry, Department of Lands and Forests, Administration Division Papers, 'O' Correspondence, File 0124854 Alcan General (1), Melrose to R. E. Powell, president, Aluminum Company of Canada, June 16, 1948 (copy). Although the copy does not bear Melrose's signature, a subsequent letter from Alcan to Melrose refers to

If after such surveys and investigations have been made, and your engineering studies demonstrate that our existing laws would not economically permit further development, I shall be very glad to discuss ways and means with my colleagues, having in mind the amendment of such laws whereby such a project might be economically pursued to the mutual advantage of both our Government and your Company.

Alcan would later call for changes to provincial law and taxation policies in order to facilitate its building program and settlement scheme at minimal cost. In 1950, Kenney summarized provincial intentions toward the Alcan project: "We will facilitate it in every way possible."[17]

The Rise of a Fisheries Defense

In public and private, the provincial government had made plain its bullish campaign to bring Alcan to BC. It had favored the company over other industrial rivals, it had positioned itself to resist competing projects in Alaska, and it had offered attractive entrance terms. Now the problem was to ensure that no internal divisions in the province displaced the project or delayed it. With wide public support reflected in the press and arguments from opposition parties focusing on means, not ends, the Coalition government moved ahead without significant protest – except, of course, from fisheries interests.

Protest emerged first within the state. Since the end of the war and the completion of the fishways project at Hells Gate, members of the IPSFC and the federal Department of Fisheries had cooperated to plan a major restoration program throughout the Fraser Basin to remove natural obstacles and dams that might inhibit salmon migration. Fisheries officials and scientists saw the Fraser as the possible antithesis of the Columbia River. Whereas dams on the Columbia were increasingly cutting into the reproduction of salmon, on the Fraser scientists seemingly had free rein to enhance salmon habitat and curb past developments. They sought

his views expressed in this letter. Duncan Campbell, in *Global Mission*, Vol. II, pp. 58–9, refers to the author as Kenney and cites an official letter in Alcan's possession.
17. BCER CF, "BC Hopes for Alcan Deal Soon," *Vancouver Sun*, December 19, 1950.

to elevate the importance of the river as a salmon spawning habitat and became increasingly involved in organizing the fishing industry to resist power-development proposals. To this group of officials and scientists, still basking in public commendation for their activities at Hells Gate, the announcement of plans to develop power in the upper Fraser for aluminum came as an unpleasant reminder of postwar realities.

What effect development would have, or the extent to which it might be modified, remained unknown. Both sites on Chilko Lake and the Nechako River mentioned in the press invited concern as important salmon spawning areas. But they remained little studied. Before the war, the Provincial Fisheries Commission had conducted occasional spawning bed surveys at these sites, and in the course of the Hells Gate investigations, the IPSFC had attempted to determine spawning returns to various sections of the basin, including Chilko Lake. However, the size of the runs and the effects that development might have on them remained in the realm of educated guesswork. Fisheries scientists did know that Chilko Lake was highly productive; calculations conducted by the Department of Fisheries suggested that up to three-quarters of the Fraser's sockeye spawned there on two of the four-year cycles.[18] If water were diverted from the lake, or if it were dammed, the effects on the salmon fisheries and recent restoration efforts could be severe. As of 1948, however, no official plans had been released, so consideration of the consequences could proceed in only an abstract way.

A growing sense of apprehension among fisheries officials spilled into public debate. After repeated attempts to gain information on the planned development and the nature of the surveys, IPSFC Chair and federal Liberal MP Tom Reid decried the activities of the "power-minded" provincial government before the House of Commons in Ottawa. The federal Liberal government must act, he insisted, to protect the fisheries, under federal jurisdiction, by strengthening existing legislation.[19] Reid's comments probably came as little surprise to the provincial government, but it did

18. NAC, RG 23, Vol 1823, File 726-11-6[7], "Memorandum on Negotiations Involving the Department of Fisheries in Connection with Developments by the Aluminum Company of Canada Limited in British Columbia," p. 3.
19. *Debates of the House of Commons*, Vol. VI (1948): 5865–7.

disturb Alcan officials. Having realized the potential conflict with fisheries concerns that a water-development project might raise, Alcan officials had already sought the counsel of H. R. Macmillan, a renowned leader in the forest industry and former president of BC Packers, the largest fish-processing firm in the province. On his advice, they had pursued a low-key approach toward the fisheries question. Rather than discuss the problem in public, Alcan asked the provincial government to allay the fishing industry's fears on its behalf. When Reid's comments hit the press and gained wide national attention, Alcan officials put pressure on the provincial government to remind officials of their role.[20] Provincial politicians acted cautiously and explained to their federal counterparts and the press that no consultation needed to be offered to the fishing industry or the federal Department of Fisheries at the present time because no projects yet existed.[21] The province was acting out a carefully orchestrated public relations program to diminish fisheries protest before it could materialize, and it was doing so by offering as little information as possible.

In light of the province's prodevelopment position, the federal Department of Fisheries followed Reid's advice and began to explore its legal powers. The authority of the federal Department of Fisheries in water management issues was not extensive. Section 20 of the Fisheries Act gave the federal department the power to insist on remedial activities caused by river obstructions. This authority, a vestige of earlier legislation that focused on the problem of mill dams, was not suited to the task of modern salmon conservation: It could be implemented only *after* a dam had been erected, and most modern dams were so large that even in the most optimistic case remedial works would only save a small portion of a salmon run.

20. BC Crown Lands Registry, Department of Lands 'O' Files, File 0124854, Dubose to Melrose, June 30, 1948.
21. BCARS, GR 1222, Premiers' Papers, Box 40, File 9, George Pearson, commissioner of fisheries, to Premier Hart, February 13, 1943. Box 165, File 10, Premier Hart to Davis, comptroller of water rights, June 9, 1943 (copy); George Alexander, assistant commissioner of fisheries; to premier, June 7, 1943; premier to Davis, May 11, 1943 (copy); Tom Reid, chairman of IPSFC, to premier, May 13, 1943; premier to Reid, May 28, 1943 (copy); Reid to premier, March 12, 1943; premier to Reid, March 18, 1943. NAC, RG 23, Vol. 1822, File 726-11-6, part 1, Premier Johnson to Mayhew, September 8, 1949; E. T. Kenney, minister of lands and forests, to Mayhew, September 15, 1949; Mayhew to Kenney, September 23, 1949 (copy).

Although revisions to the Fisheries Act in 1932 helped to clarify how fishways should be constructed and established fines for non-compliance, the legislation remained remedial, rather than proactive, in focus. According to the reading of the legislation that the Department of Justice supplied in 1948, the Department of Fisheries could rest assured that its authority extended to hydroelectric dams (although this was not specified in the legislation) and that it could demand access to plans of dams in advance of construction so that remedial efforts could be devised before projects imposed damages.[22] As subsequent events were to demonstrate, however, political pressures would condition formal legal authority.

While considering their legal powers, members of the Department of Fisheries began to canvas other fisheries agencies and members of the industry about the best means of protecting fish stocks. Initially they promoted a scheme to bring together different fisheries interests under one umbrella to form a committee focused on the fish–power problem and capable of operating a unified political program. The suggestion met with immediate assent from the IPSFC and the provincial Department of Fisheries.[23] However, George Alexander, the provincial deputy minister, suggested that the idea "would carry a great deal more weight" if it appeared to originate from industry rather than from bureaucrats.[24] At this time, however, the fishing industry seemed oblivious to the danger posed by hydro development. When Department of Fisheries officials asked executives of large cannery firms about forming a new fisheries protection committee, canners resisted, claiming that their industry organization, the Salmon Canners' Operating Committee, would suffice and that government, not industry, should tackle the problem of fish and dams.[25] Federal officials persisted and forced canners to concede that the matter of provincial jurisdiction over

22. NAC, RG 23, Box 1222, File 726-11-5[1], Stewart Bates, deputy minister of fisheries to deputy minister of justice, May 10, 1948 (copy); F. P. Varcoe, deputy minister of justice to deputy minister of fisheries, May 14, 1948.
23. NAC, RG 23, Box 1222, File 726-11-5[1], B. M. Brennan, director of IPSFC, to G. R. Clark, western director of fisheries, July 7, 1948.
24. NAC, RG 23, Box 1222, File 726-11-5[1], George Alexander, deputy minister of fisheries (BC) to George Clark, July 6, 1948.
25. NAC, RG 23, Box 1222, File 726-11-5[1], G. R. Clark to J. Macdonald, secretary of the Salmon Canners' Operating Committee, June 26, 1948 (copy); P. E. Paulson, Canadian Fishing Company, to Clark, October 21, 1948.

waterpowers complicated the problem; the federal power was limited, and pressure would have to be brought to bear directly on the provincial government, preferably with a strong showing from the industry.[26]

The committee did not form as the Department of Fisheries had hoped, but the Salmon Canner's Operating Committee was prompted to press for provincial legal protection for the fisheries. Following the advice of George Clark, the Department of Fisheries' western superintendent, Ed Paulson of the Canadian Fishing Company led a group of canners' representatives, along with the United Fishermen and Allied Workers' Union (UFAWU) and the United Fisherman's Cooperative Association, in developing an industry brief to the provincial cabinet calling for assurances that the fisheries would be protected in the event of hydro development. The brief set out the importance of the fishing industry to the provincial economy and called for amendments to the provincial Fisheries Act to grant new powers of approval to the provincial Department of Fisheries over any water development.[27] This would allay the province's jurisdictional qualms but hand fisheries interests a form of insurance against destructive power developments.[28] Although Paulson expressed his pleasure with the "initial barrage" represented by the brief when he and a number of labor representatives presented it to the provincial cabinet in January 1949, he did admit that the group "locked horns" with Minister of Lands and Forests E. T. Kenney, who was not inclined to accept their criticisms. While waiting for a response to the brief, Paulson admitted to Clark that the fish vs. power question had become "red hot" in the province: "There is no doubt that the fishing industry, which of course includes your own Department as well as the commercial aspects of it, is faced with a critical period in its existence and our steps at this time must be made with a firm foundation under them, and,

26. NAC, RG 23, Box 1222, File 726-11-5[1], Clark to Paulson, October 27, 1948 (copy); Paulson to Clark, November 10, 1948; Clark to Paulson, November 15, 1948; Paulson to Clark, November 17, 1948; Clark to Paulson, November 22, 1948.
27. NAC, RG 23, Box 1222, File 726-11-5[1], Salmon Canners' Operating Committee, United Fisherman and Allied Workers' Union, and the United Fisherman's Co-operative Association, "Submission to the Executive Council by the Fishing Industry on the Matter of the Utilization of Fresh Water Resources in British Columbia," January 4, 1949. This brief is attached to Paulson to Clark, January 6, 1949.
28. NAC, RG 23, Box 1222, File 726-11-5[1], Paulson to Clark, January 6, 1949.

naturally, with a full realization of the economic future of the country as a whole."[29] Paulson's statement reflected the extent to which the act of preparing the brief had helped to politicize the industry and to develop a sense of cooperation between the federal department and its industrial constituency. A fisheries protest, initiated within the state, was gaining the appearance of a broader coalition of interests.

The provincial response to the fisheries brief demonstrated how the cabinet balanced priorities between the existing fishing industry and the promise of power and aluminum. Instead of agreeing to the proposed amendments of the provincial Fisheries Act, the cabinet suggested only a change to the Water Act to include "commercial fisherman or fish cannery operator" in the list of interests who could legally object to the disposal of a water license and receive a hearing before the water comptroller as a result.[30] This change followed the suggestion of W. C. Mainwaring, vice president of BC Electric, who argued that it would be better to make changes to the Water Act than to enhance the authority of the provincial Fisheries Department.[31] Canner Ed Paulson found it "ridiculous" that the province was prepared to compromise only to this extent, and George Clark, a federal fisheries official, judged the amendment "of little use."[32]

The minor changes to the Water Act provided a stark contrast to provincial legislation introduced in the same session to provide a legal basis for Alcan's aluminum project. Embodied in the Industrial Development Act (IDA), the new legislation empowered the government to facilitate Alcan's interests in significant ways. The provincial government gained the authority to override normal procedures in the Water Act regarding the terms of sale, lease, or rental of crown lands and waterpowers. This allowed the province to sell water rights to the company at the lowest possible rates. Further, the legislation established a new class of municipal governance in

29. NAC, RG 23, Box 1222, File 726-11-5[1], Paulson to Clark, January 20, 1949.
30. NAC, RG 23, Box 1222, File 726-11-5[2], Paulson to Clark, February 7, 1949; The amended act followed different wording than this quotation but carried the same intent: Bill No. 38,1949, "An Act to Amend the 'Water Act.'"
31. BC Water Management Branch, Department of Lands 'O' Files, File 5254, W. C. Mainwaring, vice-president, BC Electric, to E. T. Kenney, March 2, 1949.
32. NAC, RG 23, 1222, File 726-11-5[2], Paulson to Clark, February 7, 1949; Clark to Paulson, March 29, 1949 (copy).

the province with the invention of so-called industrial townships. This provision provided Alcan with considerable authority to plan the proposed settlements, limit local taxation of the corporation, and shape the form of municipal administration.[33]

Fisheries interests saw the IDA as a defeat, sparking a call for renewed vigilance. Ed Paulson and George Clark complained privately to one another that the IDA would make the fisheries conservation aspects of the BC Water Act null and void because the third clause of the IDA granted authority "notwithstanding any law to the contrary."[34] Although this outcome might have ended the protest as practically useless, Paulson could still draw inspiration from the failure: "We [the industry] are now prepared to go right after the thing and exert every possible bit of pressure we can muster."[35] He proceeded to get in touch with labor and sports groups and received assurances of editorial support from a number of newspapers.[36] In under a year, the industry had gone from apathy to mobilization. Before the problem was concluded, however, "going after the thing" would come to mean something less public and more conciliatory than Paulson was yet prepared to imagine.

In conjunction with the federal Department of Fisheries' attempts to strengthen its legal position and organize the industry, the IPSFC launched its own scientific and political program. Surveys at Chilko Lake conducted by IPSFC scientific staff in cooperation with federal scientists with the Fisheries Research Board assessed the state of existing salmon stocks and forecast possible outcomes of dam development.[37] Reports produced from these exercises provided the Department of Fisheries with the substantive data on fisheries stocks from which to bargain with Alcan about remedial

---

33. Industrial Development Act, 1949.
34. Ibid. Clark's suspicions proved to be correct. When in 1951, Clark called on Water Comptroller E. Tredcroft to assist the federal Department of Fisheries in ordering Alcan to comply with Section 20 remedial work, Tredcroft made this reply: "I have been advised by our Legal Department that under the provisions of the Industrial Development Act of 1949 and the agreement between the Aluminum Company of Canada and the Government of British Columbia that I have no jurisdiction to order any fish protection devices." BC Water Management Branch, Department of Lands 'O' File correspondence, Tredcroft to Clark, October 1, 1951.
35. Ibid.
36. This may account for the favorable editorial regarding the brief published in the *Vancouver News Herald*: BCER CF, "Hydro and Fisheries," November 22, 1949.
37. NAC, RG 23, Vol. 1823, File 726-11-6 [7], "Memorandum on Negotiations Involving the Department of Fisheries in Connection with Developments by the Aluminum Company of Canada Limited in British Columbia," p. 12.

requirements.[38] Although important for this reason, these surveys were seriously hampered by the lack of cooperation from Alcan; the company would reveal neither its favored development sites nor the probable dam design. IPSFC scientists were thus left unable to study the relevant problems, except in a general way. Before making a formal request for water rights on the Nechako River in public hearings before the water comptroller, McNeely Dubose, Alcan's vice president and point person on the BC project, met with IPSFC staff and intimated that a dam site on the Nechako might be chosen. When Milo Bell, an IPSFC engineer, raised the possible effects on salmon, Dubose invited him to consider the project instead as a kind of "laboratory" for the fish–power problem. When Bell replied that there were many alternative sites to compare, Dubose reminded him that the Nechako site was not ensured; Chilko Lake might be the eventual location. The ambiguity also served as a veiled warning: The scientists should assist Alcan by endorsing the Nechako site or risk greater damages at Chilko Lake. Bell ended his report of the meeting with this telling observation: "One could... gain the impression that they do not care to talk about any remedial steps until they have gained their full water rights. This would make future negotiations extremely difficult as once the right is granted we would be asking for charity unless our reasonable requests are protected by limitations in their first rights."[39]

Despite the difficulties of scientific assessment, the IPSFC engaged in a variety of political tactics to weaken the Alcan proposal. Using their American representation, IPSFC commissioners attempted to draw in the U.S. federal government, claiming that the Alcan development would contradict the conservationist terms of the Pacific Salmon Convention. The U.S. State Department did not appear willing to intervene in a public manner at this early stage, but U.S. Consul Paul Meyer visited Premier Johnson in September 1949 and reminded him of the obligations contained in the Salmon Convention, expressing concerns about the possible aluminum development at Chilko Lake.[40] Apart from this meeting, however,

38. These studies are summarized in IPSFC, *A Review of the Sockeye Salmon Problems.*
39. NAC, RG 23, Vol. 1822, File 726-11-6, part 1, Milo Bell, IPSFC, to G. R. Clark, October 14, 1949.
40. BCARS, GR 1222, Box 114, File 7, note recording visit of Paul Meyer, U.S. consul to Premier Johnson, September 2, 1949.

the U.S. presence remained studiously unobtrusive: The State Department tracked the development and received reports from the Canadian Departments of External Affairs and Fisheries, as well as the IPSFC.[41] In the event of an Alcan development, however, the State Department made clear that it would abide by the actions and judgment of the Canadian Department of Fisheries, as American interests were judged to be the same as those of the federal department.[42]

Beyond the IPSFC's diplomatic initiatives, the commission's chairman, Tom Reid, launched a vitriolic public campaign against Alcan starting in the summer of 1948 and extending into the early 1950s.[43] Besides raising the Alcan problem in the House of Commons, Reid lashed out at provincial politicians for sacrificing BC's salmon industry to aluminum and attempted to portray Alcan as a self-serving "cartel" with little concern for BC's economic well-being.[44] In an attempt to awaken fears of American domination, Reid, although chair of a Canada–U.S. commission, claimed that Alcan's real intention was to export power to the United States. He also argued that insufficient public scrutiny would lead to a sellout of BC's resources and irreparable damages to the province's salmon "heritage."[45] Reid went so far as to challenge Premier Johnson to a public debate on the Alcan project in 1950 and received invitations from the provincial CCF party to lecture the provincial legislature on the matter.[46] Reid's wide-ranging critique, in short, attempted

41. NAC, RG 23, Vol. 1822, File 726-11-6, part 1, under-secretary of state for external affairs to deputy minister of fisheries, August 22, 1949; under-secretary of state for external affairs to Stewart Bates, deputy minister of fisheries, September 13, 1949.

42. NAC, RG 23, Vol. 1822, File 726-11-6, part 2, G. E. Cox, secretary, Canadian Embassy, Washington, D.C., to A. R. Menzies, American and Far Eastern Division, Department of External Affairs, January 14, 1950.

43. This note refers to general articles reflecting Reid's views. Subsequent notes refer to specific quotations: BCER CF, "Hydro Plant Called Danger to Fisheries," *Province*, June 25, 1948; "Tom Reid Denies Opposition to Alcan," *Vancouver Sun*, June 24, 1950; "Senator Reid Still Sure Plan Will Hit Salmon," *Victoria Times*, June 5, 1950; "Senator Continues Aluminum Protest," *Columbian*, April 6, 1950; "Kitimat Development 'Threat to Salmon,'" *Vancouver News Herald*, May 22, 1952.

44. BCER CF, *Vancouver Sun*, November 18, 1950, untitled.

45. BCER CF, "BC Heritage Laid at Stake in Hydro Plan," *Victoria Colonist*, November 8, 1949.

46. BCER CF, "Premier Challenged to Debate," *Province*, May 23, 1950; "Johnson–Reid Feud," *Columbian*, May 19, 1950; "Johnson and Winch Wrangle Over Reid," *Columbian*, March 7, 1950; "Winch Fails Again in Alcan Attack," *Vancouver Sun* March 7, 1950.

to reconfigure the nature of the debate over Alcan by pressing its boundaries and invoking the language of populist protest (anticapitalist, antistate, anti-American).

By his efforts, Reid did succeed in raising the profile of the Alcan case, but he also became a public foil for prodevelopment forces – the extreme voice to be discredited rather than engaged. Alcan executives fumed about Reid in private, and editorialists claimed that he was hindering development and discouraging investment in the province.[47] Only two metropolitan dailies published editorials that could be interpreted as tolerant of, or encouraging about, Reid's position.[48] Provincial politicians in the Coalition government, and particularly its Liberal elements, attacked Reid: Premier Johnson suggested that Reid's opposition might be "the final straw" to convince Alcan to depart; E. T. Kenney, the minister of lands and forests, argued that Alcan might just as easily develop power in Alaska and that Reid ought to fall in line.[49] This tough talk masked the fact, known at least to Kenney, that Alcan was firmly committed to the BC project by 1949: Dubose had assured Kenney's deputy minister of this fact in private correspondence and asked the minister to discount claims to the contrary.[50] The tone of political discourse became so heated that even some of Reid's erstwhile supporters in the fishing industry avoided being seen with him in public.[51] The federal minister of fisheries, Robert Mayhew, also disavowed Reid's views in private to the premier.[52] E. T. Kenney labeled Reid BC's "CCF Senator" – a slur from a provincial Liberal

47. BCER CF, "Not Very Helpful," *Victoria Times*, November 14, 1949; "Senator Reid's Strange Stand," *Vancouver Sun*, November 8, 1949; "BC Can Have Both," *Vancouver News Herald*, January 19, 1949; "Hindering His Province," *Victoria Times*, May 26, 1950; "Stick to Your Bagpipes Tome," *Victoria Times*, March 7, 1950.
48. BCER CF, "Reid Again Attacks Aluminum Hydro Deal," *Columbian*, November 8, 1949; "Senator Can Have His Say," *Columbian* October 29, 1949; "Hydro and Salmon," *Columbian*, January 21, 1949; "All to the Good," *Vancouver News Herald*, May 26, 1950.
49. BCER CF, "Sounding a Sour Note," *Victoria Times*, May 18, 1950; "Reid Menace to Alcan Says Premier," *Vancouver Sun*, May 17, 1950; "Premier Hits Back at Senator," *Province*, March 2, 1950; "Kenney Hits Criticism of Alcan," *Province*, November 12, 1949.
50. BC Crown Lands Registry, Department of Lands 'O' Files, File 0124854, McNeely Dubose to George Melrose, March 14, 1949.
51. NAC, RG 23, Vol. 1822, File 726-11-6, part 2, Ed Paulson, general manager, J. H. Todd & Sons, Ltd., to Dr. A. L. Pritchard, Department of Fisheries, November 8, 1949.
52. BCARS, GR 1222, Premiers' Papers, Box 114, File 7, Mayhew to Premier Johnson, November 9, 1949.

politician that equated Reid with the Left and as an opponent of continental security measures.[53]

Although part of the intensity of this reaction may be explained by the personalities involved and the sensitivities of BC politicians to party loyalties at a time when the coalition government was undergoing internal strains, the debate also suggests the importance of the Cold War context. In 1950 war had commenced in Korea and Canada was involved. The Pacific focus of the war and the perceived threat of communism thrust BC into a new position as discussions of international involvement and continental security developed. To some extent, the Alcan project became symbolic of BC and Canada's involvement. Alcan placed the project on an advanced schedule to respond to new military demands. As construction proceeded, plans included a deep granite bunker for the powerhouse and penstocks were buried within a mountain, the better to protect against air attacks. Alcan officials and C. D. Howe, then Canadian minister of trade, led a delegation to the U.S. government to attempt to negotiate contracts for the new aluminum project.[54] At U.S. House of Representatives hearings on aluminum supplies in 1951, U.S. producers expressed resentment about the Canadian project, but were countered by U.S. State Department officials who argued that new aluminum supplies were needed and underlined the insurance that the Alcan project would provide to the United States.[55] "When we needed aluminum from Canada [during the war], we got it," said one member of the State Department, adding there was no reason to believe this would change.[56] Editorialists in BC railed against the perceived slights toward Alcan and Canada uttered at the hearings and called on BC politicians to explain that BC "won't be treated like a second-rate Central American banana state.... This continent is supposed to be converted into the arsenal of democracy. Then let's start producing defence materials where

53. BCER CF, "Reid Rapped on Aluminum," *Vancouver Sun*, December 5, 1950; "Raps Reid for Hydro Statement," *Columbian*, January 22, 1949.
54. BCER CF, "Cabinet Gives Nod to BC Alcan Plant," *Vancouver Sun*, December 22, 1950.
55. BCER CF, "BC Aluminum Project Meets U.S. Opposition," *Vancouver News Herald*, March 7, 1951; UBC, Doyle Henry, Papers, "Alcan Hasn't Negotiated for U.S. Aid in British Columbia Project, President of Parent Firm Asserts," *Wall Street Journal*, January 30, 1951.
56. UBC, Doyle, Henry Papers, "Alcan Hasn't Negotiated for U.S. Aid in British Columbia Project, President of Parent Firm Asserts," *Wall Street Journal*, January 30, 1951.

they can be produced best and forget about picayune commercial rivalries."[57] No government contracts would be signed in the end, but Alcan would supply ingots to U.S. producers throughout the Korean War.[58] In the context of these discussions and the background of military buildup, Tom Reid's complaints could be read not as comments about fish and power but as an intervention into a broader spectrum of ideological and security issues. This was probably not his intention, but the critic could not control the context in which his ideas were received.

Even before the pressures created by the Korean War context arose, powerful segments of the fishing industry decided to abandon Reid's lead and to bargain with Alcan. Having failed to ensure fisheries protection through legislation, and convinced that the provincial government would favor aluminum over salmon, the federal minister and major canning interests attempted a pragmatic approach. The presence of two possible sites – Chilko Lake and the Nechako River – provided the fortuitous possibility of a horse trade: The more important salmon spawning grounds could be exchanged for tacit acceptance of development on the less important northern site. Although the details are admittedly obscure, there is evidence that in the fall of 1949 members of the Fraser canning industry sought to develop a private agreement with Alcan to save the spawning grounds at Chilko Lake. Fisheries Minister Robert Mayhew knew of this plan and may well have played a role in initiating it. James Eckmann of the Canadian Fishing Company wrote to him in late September 1949, referring to previous conversations on the matter. Eckmann reported that[59]

The industry held a meeting Monday afternoon to discuss the question of the water application for the Aluminum Co. and without mentioning the conversation I had with you, a discussion took place and finally a recommendation was agreed upon whereby the industry would make no protest or oppose Alcan's two applications for water rights on the Nechako River but it would be distinctly understood that they would withdraw the application insofar as the Chilko is concerned.

57. BCER CF, "Retaliate!" *Vancouver Sun*, January 27, 1951.
58. Litvak and Maule, "Alcan Aluminum Ltd.," pp. 51–3.
59. NAC, RG 23, Vol. 1822, File 726-11-6, part 1, James Eckmann, president of the Canadian Fishing Company, to Mayhew, October 25, 1949.

Later, Eckmann and J. M. Buchanan of BC Packers Ltd. visited Victoria with the intention of presenting the substance of this compromise to the premier and the minister of lands and forests, E. T. Kenney.[60] Two years later McNeely Dubose referred to this private agreement when resisting federal calls for remedial work.[61] It is impossible to say who knew of this reported trade-off. Logically, the Salmon Canner's Operating Committee could have sponsored the meeting of "industry" described by Eckmann, but it is difficult to say. The point is that members of the fishing industry tried to forestall an unfavorable outcome by giving up on the Nechako site under the threat of a worse outcome. They stood outside the bounds of public debate and sought a cruder political bargain.

By the time a water licensing hearing for Alcan's application was called in Victoria, October 31, 1949, little had been revealed about the desired development location.[62] The Chilko site, on which the IPSFC had prepared substantial scientific evidence, vanished from the agenda in favor of the northern Nechako site, about which much less was known. A number of engineering and site planning considerations shaped Alcan's decision, but so too did the threat of legal and political problems associated with the fisheries question at Chilko Lake. Although the IPSFC objected to the Nechako project on the grounds of possible effects on the fisheries, most representations to the provincial water comptroller acknowledged that development would have to occur somewhere, but suggested it would be helpful to conduct some kind of preliminary studies on the

60. UBC, Doyle, Henry Papers, Box 1, File 1-4, Buchanan to Doyle, November 2, 1949. Buchanan refers to the meeting in this letter and says that their presentation to the premier and minister paralleled their later statements to the water comptroller in the water license hearing. By the time of the hearing, of course, the position may well have changed the bargaining point of the province and Alcan.
61. NAC, RG 23, Vol. 1823, File 726-11-6, part 5, "Meeting with Officials of the Aluminum Company of Canada," August 9, 1951. The statement reads: "Mr. Dubose referred to a statement credited to him to the effect that the fisheries interests had intimated that if the original Chilko plan were abandoned in favour of the Nechako there would be no fisheries problem. He explained that this understanding was very definitely given to him by a group purporting to represent the BC Salmon Canning Industry at an interview they had with him in Victoria at the time. He wished to record however that official Government Fisheries agencies had not been associated with this statement."
62. BCARS, GR 880, Power and Special Projects, Box 4, File 1, "Record of Hearing on Applications by the Aluminum Company of Canada Ltd., for Water Licenses on the Nechako and Nanika Rivers Held Victoria, BC, October 31, 1949." Another meeting with local residents who would be affected by the project was held at Wistaria, October 24, 1949.

impact on the fisheries. To this, Alcan objected as an unnecessary delay.

Beyond the representations made on behalf of the fishing industry, there were few others present to raise questions about the project's social consequences. A representative of settlers in the upper Nechako region who would be flooded out by the reservoir noted the costs for people living in the area but also thanked Alcan for its consideration. An elderly woman rose to speak on behalf of wilderness values and read out a letter from Lady Tweedsmuir criticizing the sacrifice of wilderness to progress in areas within and bounding the park named after her husband, Lord Tweedsmuir, the former Canadian governor general. Arguably some of the most affected social interests in the coming development, Cheslatta T'en peoples, inhabiting and trapping portions of territory that would be flooded, were not present and had not yet been informed of the project or its costs. Major Farrow, the provincial water comptroller, took the views offered into consideration. In the water license granted to Alcan, he inserted one concession to the fisheries interest by insisting that Alcan should present its plans before construction to the federal Department of Fisheries.[63] For all intents and purposes the water license deal was done. Alcan had a legal right to the waters of the Nechako River and permission to flood the areas it required. It also had favorable tax and policy terms. What mattered now was moving the operational side of the development forward (see Map 6).

The Politics and Limits of Regulation

The politics of development sometimes turn on small clauses. In the coming months, as Alcan proceeded to marshal resources, construct roads, and begin dam construction on the Nechako River, it would fend off the attempts of the federal Department of Fisheries to impose remedial measures under the federal Fisheries Act. But to some extent, the battle had already been won. Provincial support for the Alcan development was unshakable. The legal powers of

63. BCARS, GR 1222, Premiers' Papers, Box 117, File 3, Farrow, BC water comptroller to Mayhew, minister of fisheries, December 12, 1949.

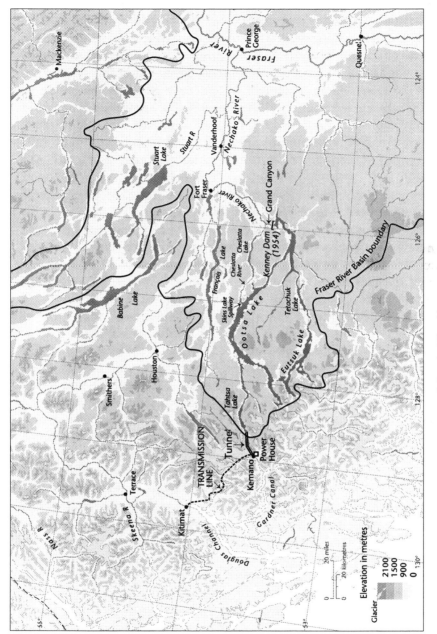

Map 6. The Alcan project.

171

the federal Department of Fisheries remained unclear and untested. The fishing industry had removed itself from the front lines of the debate to allow the scientists to take over and recommend remedial measures. Some prominent fishing interests had already conceded the Nechako site in order to remove development pressure from Chilko Lake. What remained to be determined was the extent to which Alcan would have to live up to the commitment contained in the water license. How much would it reveal? What did it need to reveal? And what, if any, changes would it be forced to make?

In the summer and fall of 1950, as Alcan began construction, scientific staff of the IPSFC and the federal Department of Fisheries surveyed the Nechako Basin with the assistance of Alcan engineers and attempted to determine possible effects on salmon. The main problem was not to evaluate the loss of spawning grounds. Some would disappear under the reservoir behind the Kenney Dam, but they could not be saved. The more serious problem was the extent and seasonal timing of the diversion. This was also a crucial aspect of the project itself. The volume of water to be extracted from the upper Fraser Basin and piped to the coast was massive; later estimates would put it at 3% of the Fraser's annual flow on average.[64] The diversion would not be uniform over the year. At different times, depending on the height of the reservoir storage, more or less water would need to be held back from the Nechako River. Probably the greatest diversions would occur in spring and summer. This mattered from a fisheries perspective for a variety of reasons. Reduced flows would affect the riverbed, raise water temperatures, make summer upriver migration more difficult, and reduce winter cover for fry. Divert a river, in other words, and there would be consequences downstream. On the lower Nechako, important salmon runs would face these problems; in the upper Fraser, runs to Stuart Lake, a major spawning site, would also be affected. As fisheries scientists weighed these possible outcomes, conducted on-site inspections, and looked over Alcan engineering plans, they developed a possible remediation strategy that would avert the worst effects of diversion. Would it be possible, they asked, to release water from the project's dam at critical points of the year to maintain lower

64. Moore, "Hydrology and Water Supply," p. 28.

water temperatures in the lower Nechako and assist salmon migration in the summer months? They believed this could be done and published a Department of Fisheries report stating their findings and estimating a total cost for remedial work at approximately $1,410,000.[65]

Alcan responded to the report a month after its release. In a meeting in Ottawa on February 1, 1951, McNeely Dubose, Alcan vice president, informed Minister of Fisheries Robert Mayhew that the remedial work was unnecessary and too expensive. Mayhew was in a difficult political position. His obligations as minister and a BC MP stood in the background, as did the knowledge that aluminum would play a role in military affairs ahead. He insisted that Alcan should expect expenses reaching $1 million and indicated that the company would not look well should damage befall the fisheries.[66]

This opening debate suggested the power of private capital to play off different levels of government against one another and to employ political pressures to diminish regulatory scrutiny. Alcan would continue to press its advantage in ensuing negotiations, placing the federal Department of Fisheries on the defensive, despite its legal powers.[67] In later discussions, Dubose was, by turns, charming, indulgent, and ornery. The shifting postures infuriated and confused Department of Fisheries staff, who spent many hours composing memoranda attempting to grasp his direction, wondering if his positions were bluffs. Dubose's performance was brilliant. The Department of Fisheries wanted cold-water releases and asked

65. NAC, RG 23, Vol. 1822, File 721-11-6[3], C. H. Clay, division engineer, to A. J. Whitmore, chief supervisor of fisheries, January 15, 1951.

66. NAC, RG 23, Vol. 1822, File 721-11-6[3], A. L. Pritchard, "Memorandum re Meeting February 1, 1951 to discuss Aluminum Company of Canada Development in British Columbia in relation to Fisheries," February 2, 1951; Vol. 1823, File 726-11-6 [7], "Memorandum on Negotiations Involving the Department of Fisheries in Connection with Developments by the Aluminum Company of Canada Limited in British Columbia," March 10, 1952; further aspects of the negotiations are covered in File 721-11-6[5], Mayhew to Dubose, October 15, 1951; Dubose to Mayhew, September 26, 1951; Mayhew to R. E. Powell, president, Aluminum Company of Canada, September 17, 1951.

67. Some of the other meetings are reported in NAC, RG 23, Vol. 1822, File 726-11-6[3], C. H. Clay to Whitmore, March 7, 1951, regarding meeting with Alcan engineer Kendrick; File 726-11-6[4], S.B. [Stewart Bates], "Memorandum to the Minister, Re: Discussions with Mr. Dubose RE Aluminum Company of Canada," April 16, 1951; "Memo for File June 13, 1951," regarding meeting among minister, President Powell, Aluminum Co., Dubose and C. D. Howe.

for a low opening in the dam for the purpose. Dubose dithered, raised the safety risk (the dam might break: "I am not exaggerating"[68]), and questioned the biological foundations of the department's advice. One fisheries department scientist summarized one of Dubose's pugnacious letters in three lines: "We do not think you need cold water. In any event, we do not intend to give it to you. We know the weakness of your arguments and we intend to stress them."[69] The department finally settled for no cold-water releases.

Further confusion surrounded the release of water to allow for salmon migration in the summer and cover in the winter. Dubose questioned the Department of Fisheries figures on the required releases, hired expert consultants to criticize the department's findings, and then offered to release water at a far lower level ($100 \text{ ft}^3/\text{s}$) than the department desired. Although the department continued to argue the point, Dubose succeeded in ending discussion of the possibility of major releases from Alcan's main project reservoir by suggesting a secondary spillway via Cheslatta Lake. Even when this point was apparently conceded, however, Dubose continued to press: seeking funds from the department to assist construction and backing off earlier commitments as to the design of storage works.[70] Before agreement was finally reached, Department of Fisheries scientist A. L. Pritchard claimed in a memorandum to his department, "We have compromised and accepted maximum risk. The Company has done nothing."[71] In the final analysis, Alcan would do something, but it was far below the Department of Fisheries' early expectations.[72]

68. NAC, RG 23, Vol. 1823, File 726-11-6 [7], Dubose to Mayhew, March 21, 1952. Dubose denied the necessity of a passage in the dam during earlier negotiations: NAC, RG 23, Box 1823, File 726-11-6[5], "Meeting with Aluminum Company," August 14, 1951.
69. NAC, RG 23, Vol. 1823, File 726-11-6[7], A. L. Pritchard memo to deputy minister, March 21, 1952.
70. NAC, RG 23, Vol. 1822, File 726-11-6[4], G. R. Clark, "Memorandum to Minister," April 16, 1951. On Dubose's continuing negotiating tactics, see RG 23, Vol. 1823, File 726-11-6[8], "Supplementary Memorandum No. 1 on Negotiations Involving the Department of Fisheries in Connection with Developments by the Aluminum Company of Canada Limited in British Columbia (Covering Period from March 10, 1952 to January 21, 1953)."
71. NAC, RG 23, Vol. 1823, File 726-11-6 [7], A. L. Pritchard, "Memorandum to the Deputy Minister, Relevant Factors for Consideration in Future Negotiations with the Aluminum Company of Canada Ltd," March 10, 1952.
72. NAC, RG 23, Vol. 1823, File 726-11-6, Dubose to Mayhew, March 27, 1952; Mayhew to Dubose, March 28, 1952.

To correct for the effects of one dam, another would be necessary. To provide the necessary water for migrating sockeye in the summer months, Alcan and the federal Department of Fisheries agreed to create an additional spillway and reservoir on the Cheslatta system, tributary to the Nechako. This would provide the necessary extra storage to be released at appropriate times without diminishing the company's power supply. After this project gained the agreement of both the Department of Fisheries and Alcan, it was rushed to completion.[73] With a view to avoiding effects on the first runs to bear the consequences of upstream development, both the provincial and federal governments pressed forward on approving the necessary regulations. Through fortuitous weather conditions, this approach appeared entirely satisfactory in the first few years after construction, when high water levels canceled out potential problems. One fisheries scientist suggested that this result was bad luck: It discredited fisheries criticism and left the problem unsolved. Agreeing in future years on the quantities of water to be released remained contentious.[74]

All of these actions were taken with a view to conserving the fisheries, but they paid little regard to the social effects of further flooding. Already the Alcan project had been responsible for the displacement of settlers on the margins of Ootsa and Eutsuk Lakes. With the power to expropriate these parcels of land, the company had proceeded to negotiate over terms with settlers.[75] The more dire consequences came for native peoples in the region. Over 200 Cheslatta T'en people had lived and trapped in the area to be flooded by the new Cheslatta Dam. When waters rose in April 1952, they were asked by Department of Indian Affairs officials to abandon the territory and sign papers, which gave Alcan legal title. They did so without a full understanding of the consequences or third-party advice. In the process, they lost valuable trapping territories and witnessed ancestral burial grounds disappearing under water. During the final phases of flooding, a number of native families who

73. NAC, RG 23, Vol. 1823, File 726-11-6 [7] Mayhew to W. E. Harris, minister of citizenship and immigration, March 28, 1952; and C. H. Clay, "Memorandum," March 12, 1952.
74. See ongoing correspondence in NAC, RG 23, Vol. 1823, File 726-11-6.
75. BCER CF, "Farmers Ask Aluminum Co Compensation," *Vancouver Sun*, October 26, 1949; "Alcan Promises Compensation," *Province*, October 25, 1949; "Alcan Buying Tactics Assailed," *Victoria Colonist*, October 28, 1951.

had not been informed of the situation were evacuated.[76]After an initial period of confusion and displacement, the Cheslatta T'en received several small and widely dispersed land parcels around the region as compensation. These areas did not become reserves until 1964. The former proximity of Cheslatta T'en peoples was broken. By the early 1990s the treatment of the Cheslatta T'en in this process would lead to substantial settlements from the federal government. The controversy surrounding the implementation of the project nevertheless continues.[77]

Conclusion

Numerous helicopters clattered over the Nechako project as it came to completion in the early 1950s. Passengers had very different views from those seen by Alcan executives touring the region in 1948. From a distant region, surrounded by trees and dispersed settlement, lakes, and a large river, the Nechako and upper lakes area had been transformed into an industrial landscape. The final project dammed the Nechako River with a massive rock-filled, clay-core dam that raised water levels 300 ft at its face. The dam flooded several lakes into one large circular reservoir chain to a western extremity on Tahtsa Lake. From this point, an intake diverted water to the coast through a pipeline that entered a powerhouse near the outlet of the Kemano River. The powerhouse was located within the housing of a granite mountain to protect against air assaults. Transmission lines ran from the powerhouse over 50 miles of rough terrain to Kitimat on the Douglas Channel. The Kitimat site provided suitable land for a smelter and offered ocean access to incoming freighters carrying bauxite and alumina from Jamaica and for outgoing ships carrying aluminum ingots. Alcan constructed a model industrial town, planned by a New York design firm, and immigrant workers arrived to run the facility. By the late 1950s, Alcan would estimate a total investment in the project

76. NAC, RG 23, Vol. 1823, File 726-11-6[8], A. J. Whitmore, "Memorandum, Re: Murray Lake Dam – Cheslatta System," April 29, 1952. Whitmore reported the evacuation of nine Indian families and the removal of Indian graves.
77. Buhler, "Come Hell and High Water"; Rankin and Finlay, "Alcan's Kemano Completion Project"; *Looking Forward, Looking Back*, pp. 475–82.

of $440 million. By 1959, the project had an installed capacity of 1,050,000 hp. The scale of this project was unprecedented in the BC context. No industrial development in the province's recent past could match it for size or importance.[78]

Journalists who monitored the project's construction celebrated the contrasts between industrial development and seeming wilderness. Concrete and steel read against rock, forest, and river suggested dramatic shifts, but not ruptures. The contrasts, rather, pointed to the promise of a new middle landscape, where British Columbians would forge a prosperous future out of an industrialized nature where industry and the natural world would accommodate one another. Paul St. Pierre imagined the Nechako dam and its future, where "A lone man, a watchman, will remain . . . while nature starts the slow process of covering the grey gashes in the mountainsides with green forest."[79] S. W. Fairweather, the vice president of research and development for the Canadian National Railways suggested that the Alcan project proved that humans were more efficient than beavers or ants. "We are changing the course of rivers, creating lakes where none existed before, making electric energy inside solid rock, drilling tunnels through ten miles of mountain. If we can work with people as well as we can work with nature, I have no fear for our future in the sterner realm of world affairs."[80] Even as the Alcan project readied to produce aluminum ingots for military sale, Fairweather believed that it stood as a monument to future peace: The project's evidence of ingenuity in nature could surely be transferred to society. The power of aluminum appeared protean to these observers. Physical energy suggested human energy. Industry blended into nature. Growth and change augured improvement and provided comparative advantages.

But when would the advantage of waterpower outweigh others in the province? In the Alcan case, the provincial government's position was clear: A new, large industrial development held priority in resource decisions over all other interests. To judge by some of

78. Aluminum Company of Canada, "Kitimat-Kemano": Harry Jomini, "The Kenney Dam."
79. BCER CF, Paul St. Pierre, "Alcan Dam to Be Lonely Monument," *Vancouver Sun*, June 6, 1952.
80. BCER CF, "Manifestation of Confidence Says Researcher of Kitimat," *Columbian*, June 10, 1952.

the political rhetoric of these years, the position was more general than that. With major expansion plans driving forward the BC Electric system in the province's cities and steady growth under the BCPC in the provincial hinterlands, some power promoters argued that hydroelectricity offered dynamic possibilities for BC and deserved support and priority in all fields of resource development. Against that view, however, stood an increasingly unified fisheries defense, linked to various government departments and industrial constituencies, who felt anxious at the possibilities of further development and its effects on the Fraser's salmon. How would fish and power play out in future resource contests? Did the Alcan case set a precedent and, if so, of what kind?

6

Fish versus Power

In October 1957, BC Premier W. A. C. Bennett called a press conference to deliver the "most momentous announcement" of his life. The Peace River, he said with excitement, would be dammed. Bennett was a populist, a regionalist, and a promoter of BC's resource wealth. After an early career as a hardware store owner in a small interior town, he had risen through provincial politics to lead the province under a new right-of-center party, Social Credit, beginning in 1952. One of his earliest and most enduring goals was to use the provincial state to open the Interior and North to development. His government invested heavily in infrastructure projects to overcome the barriers of distance. He made overtures to international capital to develop the province's natural resource wealth. By the mid-1950s, in a period of buoyant economic growth, some of those efforts began to bear fruit. The Wenner-Gren corporation, representing the interests of Axel Wenner-Gren, a Swedish multimillionaire and notorious World War II arms dealer, applied to develop BC's Peace River region. The corporation's plans were ambitious. They included a railroad along the Rocky Mountain trench, pulp and paper and mining developments, as well as an enormous hydroelectric project on the Peace River. As surveys proceeded, hydroelectricity gained priority. Bennett extolled the possibilities. Cheap power would attract industry. Northern electricity would supply the province's needs well into the future. The Wenner-Gren project would help to fulfill the dream of connecting distant northern resources to the provincial heartland.[1]

1. "Huge Power Plan for Peace River," *Vancouver Sun*, October 9, 1957.

179

The announcement of the Peace River program ensured nothing; it was but another in a long string of promotions that the premier had learned to use in securing immediate and consequential ends. Already during the 1950s, the province had seen a number of larger-than-life development schemes fail – one, on the upper Columbia, in which the Kaiser Corporation, an American aluminum producer, would develop a storage dam in Canada in order to produce greater firm power at existing U.S. facilities downstream, had provoked the concern of the federal government. Against the wishes of Bennett and his government, the federal Liberals passed legislation canceling the project in the name of protecting resources in the national interest. The project, Bennett later suggested, had been merely a strategic lever in negotiations for Columbia River development with the United States. Whereas the United States simply wished Canada to cooperate in an integrated Columbia River program, Bennett wanted to emphasize that BC and Canada had other development options. Whatever the truth behind this claim – and it would, after all, have required considerable foresight and a high tolerance of risk – some of the key problems of jurisdiction, international negotiation, and compensation in transnational river development had received attention.[2]

Just as the Kaiser promotion had focused attention on the Columbia, so in the late 1950s did the Peace project announcement: By making the Peace the cornerstone of future power supply, the province effectively removed the immediate need for development on the Columbia, raising Canada's bargaining strength with the United States. Substantial sums would now have to be paid to gain Canada's approval for integrated Columbia River development. When completed in 1961, the Canada–U.S. Columbia River Treaty imposed obligations on, and provided benefits for, both countries. Canada was charged with building three upper-basin storage dams to benefit downstream river regulation and provide enhanced flood control (see Map 7). In exchange for foregoing other development opportunities on the upper Columbia, Canada received downstream benefits: One-half of the additional power generated by downstream projects would be returned to Canada,

---

2. Mitchell, *WAC Bennett*, p. 286; Swainson is more skeptical of this claim than Mitchell. Swainson, *Conflict Over the Columbia*, p. 65.

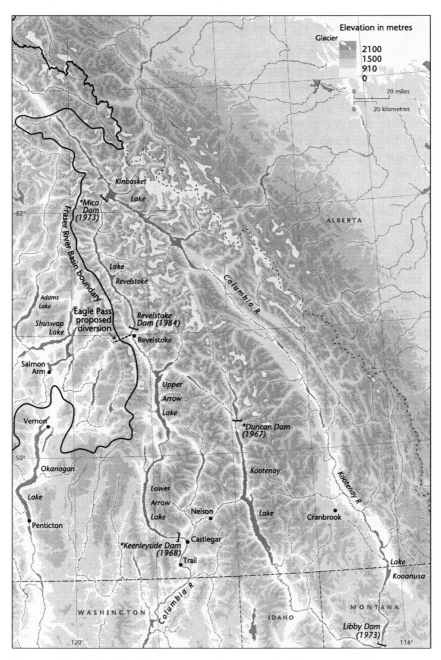

Map 7. The Columbia River projects. Dams marked by an asterisk are treaty dams.

181

as well as payments for one-half the estimated benefits from flood control. However, the federal Conservative government signed the treaty without provincial approval. Before agreeing to its terms, Bennett exacted a major concession from Ottawa: an end to the national ban on power exports. This concession would pave the way for an expanded continental electrical-energy market. Following the final ratification of the treaty in 1964, BC arranged to sell its downstream benefit power over twenty-five years to a group of U.S. utilities for $254 million, funds that would be used, in part, to construct the upper Columbia dams. At the same time, Bennett moved forward with development of the northern Peace River. When the major provincial utility, BC Electric, balked at purchasing Peace power, claiming that it would be prohibitively expensive, Bennett completed the performance by expropriating both the Peace River Development Corporation (the successor to the Wenner-Gren concern) and BC Electric and establishing a new provincial corporation, BC Hydro in 1961.[3]

Within a decade, Bennett's northern and interior development agenda had brought two large, continental rivers under development. It had also incidentally protected the Fraser River for salmon. From the perspective of the early 1950s, this outcome was surprising. In 1950, knowledgeable commentators judged the Fraser to possess the most economical power sources for the province; within a decade, the two rivers that had formerly been ranked well below the Fraser in importance would instead be developed and provide the pillars of electrical development. How that came to be so was not for lack of attempts to develop the Fraser nor the result of an inevitable political concern for conserving the fisheries.

3. The most substantial study of the Columbia Treaty remains Swainson's carefully argued *Conflict Over the Columbia*. The Wenner–Gren episode is best explained by Wedley, "The Wenner-Gren and Peace River Power Development Programs." Wedley's paper is a section of his thesis, "Infrastructure and Resources," pp. 247–310. The political events that made up the two river policy are treated in Mitchell, *WAC Bennett*; Robin, *Pillars of Profit*; Sherman, *Bennett*; Williston and Keller, *Forests, Power and Policy*, especially Chapter 2, "The Two Rivers Policy." These problems are approached from an entirely different angle (and set of sympathies) in Swettenham, *McNaughton*, Chapter 6, "The International Joint Commission." Jeremy Mouat considers these events from the perspective of West Kootenay Power in *The Business of Power*. For an analysis of the two rivers policy in the context of national and more contemporary debates, see Froschauer, *White Gold*.

More than five different hydro development schemes floated momentarily and then sank on the Fraser in the 1950s. The three most important projects – the Moran plan, a scheme to dam the Fraser north of Lytton, proposed by the Moran Development Corporation; the Columbia-to-Fraser diversion scheme, which would dam both the Fraser and Thompson Rivers, backed by General Andrew McNaughton, chair of the Canadian Section of the International Joint Commission (IJC); and the System A plan of the Fraser Basin Board that would place ten multipurpose dams in the upper Fraser Basin – each fueled widespread debate across the province and affected the course of negotiations on the Columbia and energy politics in the province more generally. How these projects arose, who supported and protested them, and what debates they produced and affected are the questions that shape this chapter.

Fundamentally, the fish vs. power debate on the Fraser helped to displace hydro development into other river basins; once major development occurred on the Peace and Columbia, the Fraser was insulated by implication. Before that time, fisheries defenders, power promoters, and politicians all understood that the future of the Fraser was up for grabs. The fish vs. power debate over the Fraser served as both an independent variable, having an impact on the context of power politics across the province, and as a dependent variable, causing particular groups to adopt positions on the Columbia and Peace questions based on their views about Fraser River affairs. In turn, the development of the Peace and Columbia Rivers had an impact on the context of the fish vs. power debate by undercutting demand for hydro development and by directly affecting rival projects on the Fraser. Bennett's approach to hydro politics focused on the Peace and the Columbia, but in the background lay the problem of the Fraser.

Besides changing the course of BC's hydro history, these developments also provided grist for British Columbians to debate the meaning of regionalism in both Canada and BC, to ponder the appropriate uses of technology, and to wonder at the powers and limits of nature. In debating the problems, British Columbians revealed their mixed impressions of the promise of development, their insecurities about past political grievances, and their anxieties about the Cold War. The fish vs. power debate turned into an open rhetorical field: A contest over resource allocation invited profound questions of nature, culture, and place.

Promoting Power

"The wise exploitation of the Fraser River," Harry Warren began, "represents one of the greatest and most thrilling hazards that lie ahead."[4] The comment from the UBC geologist signaled a philosophy of wise use – in which resources are put to utilitarian ends, with a view to perpetual exploitation – but also an infectious excitement in the challenges of BC's postwar development.

The venue for Warren's presentation was not his university classroom, but a conference. Since 1948, Warren had helped to organize the annual BC Natural Resources Conference that brought together leaders from industry, government, and academe to discuss pressing problems of BC's resource economy.[5] On this occasion, in 1952, Warren's subject was the future development of the Fraser River. For many years Warren had pushed the subject of hydropower in BC: During the war he had called on the provincial government to expand hydro facilities and peppered Alcan with encouraging advice. His concern at this stage was neither personal nor financial; he simply believed that BC had water wealth that could be exploited *and should be exploited* if the province and the West were ever to attain their proper status in Canada, North America, and the world.

Warren's talk was not the only one concerning energy and power matters in 1952, nor was it the first at these conferences to discuss the Fraser River as a power source. Only the year before, a special forum on "Fish and Power" introduced the problems attending dam development on the Fraser.[6] Speaking on behalf of power, Samuel Weston, the chief of the BCPC, H. L. Purdy, vice president of BC Electric, and a group of other prominent individuals in the field asked fisheries representatives to consider the

---

4. Warren "National and International Implications," p. 257.
5. For a useful discussion of the importance of these conferences in BC's conservation debates in the 1950s and 1960s, see Keeling, "Ecological Ideas," pp. 7–23.
6. "Forum: Fish and Power."

inevitable demand for the river's power and the economic values that hydro power might bring in comparison with the fisheries. In reply, fisheries representatives, such as Milo Bell of the IPSFC and A. H. Sager of the canners' lobby group, the Fisheries Association of BC, made every effort to distance themselves from an obstructionist position, but called on power developers to dam non-salmon-bearing streams before turning to the Fraser. From the audience, the nature writer and activist Roderick Haig-Brown, as well as McNeely Dubose, vice president of Alcan, rose to make pointed observations and queries. Haig-Brown criticized power developers for withholding information from the public about development plans; McNeely Dubose defended Alcan's integrity and insisted that his company had been transparent in the development process. Although each speaker was at pains not to dismiss the rival concern and called for a rewriting of the forum title from "Fish *or* Power" to "Fish *and* Power," the conference transactions nevertheless captured the spirit of the engagement in two photographs, showing the Fish and the Power speakers in juxtaposition. There was a confrontational tension here that talk of cooperation could not undo.

These discussions reflected the growing sense in BC that the development of the Fraser was both unavoidable and necessary. In his portrait of the Fraser River, published in 1950, journalist Bruce Hutchison put the matter tersely, "Not long will [the Fraser] remain unused."[7] In the same year, Premier Byron Johnson warned that "The time is coming when the people of the province will have to decide whether they want to develop power, or stay as they are, protecting the fishing industry."[8] The looming construction of the Alcan project and the possibility of others forced Tom Reid, chairman of the IPSFC, to declare privately that if dams on the Fraser went ahead, the Salmon Commission might as well "just fold up"; its mandate to restore the river would be impossible.[9] "A big industrial fight is shaping up in British Columbia," noted the *Victoria*

---

7. Hutchison, *The Fraser*, p. 337.
8. BCER CF, "Fisheries or Power?" *Vancouver News Herald*, May 19, 1950. The quotation appeared also in the *Vancouver Sun*, May 17, 1950, in a slightly different form.
9. NAC, Pacific Region, RG 23, Vol. 2301, Folder 1, Proceedings of IPSFC Meeting, February 2, 1951.

Times in 1950, "fish vs. power."[10] One critic of this heated rhetoric and the assumptions it masked wrote in the *Vancouver Sun* in 1949 that it was impossible to believe that dam boosters understood the threat posed to salmon or were credible in speaking on the matter: "To entertain any hope of maintaining the salmon run under these conditions," wrote engineer Paul Smith, "is to be optimistic to the point that could be justified only by conviction that help will come from supernatural agencies."[11] Another writer in the same newspaper dismissed Smith's pessimism and called for a "healthy compromise."[12] For all of the confusion about consequences, the promises of technology, and the perils of nature that this discussion contained, it turned on a sense of foreboding: a choice was coming, a challenge of self-definition.

"As a source of energy," Warren continued in his 1952 address, "the Fraser may be considered the mainspring of British Columbia."[13] Dammed on its main stem, the river could provide enormous power, captured, gearlike, in a series of integrated projects, the largest of which would be north of Lillooet at a railway siding called Moran. In a stylized drawing of the Moran concept carried in the press the day after Warren's talk, four hulking dams, thousands of times the actual size, bore down in relief on the river, technology imposing itself on nature (see Figure 3).[14] "Here is the site of power development," reveled a journalist, captured by Warren's vision, "that would surpass the St. Lawrence Seaway plan, tower over the Kitimat project like a colossus and known hydro records left and right."[15] The Moran site was the key: The dam here would stand 720 ft above the river, flood a vast area, and produce massive amounts of energy. The location had first come to light in 1934 when provincial surveyor S. H. Frame described its characteristics for the BC Water Branch.[16] Warren was taking

10. BCER CF, "Fish vs. Power Problem Soon Will Confront BC," *Victoria Times*, May 12, 1950.
11. BCER CF, "You Can't Kid a Salmon," *Vancouver Sun*, April 23, 1949.
12. BCER CF, Roy Brown, "Fish or Industrial Power? BC Can Have Both?" *Vancouver Sun*, April 25, 1949.
13. Warren, "National and International," p. 257.
14. BCER CF, *Vancouver Sun*, October 24, 1952.
15. BCER CF, "Dot on Map Possible Site for Vast BC Power Project," *Vancouver Sun*, October 24, 1952.
16. BCARS, Ad MSS 1147, S. H. Frame Papers, Box 1, 1934 diary. Frame said of Moran, "Site found to be a good one."

FUTURE FRASER RIVER development in B.C. interior could produce as much as one quarter of all electrical energy now developed in Canada. Thousands of acres of Cariboo range could be irrigated and deepsea ships could sail to Chilliwack when this happens.

Figure 3. "The Future Fraser River."
(*Source: Vancouver Sun*, October 24, 1952.)

187

the next step and imagining how the Moran could provide British Columbians with the means to realize the river as their progressive mainspring.

The project would be a mainspring rather than simply a power generator, because the entire Fraser Basin, the heartland of the province, would be reconstructed on its basis. Warren rattled off the possibilities: The reduction and stabilization of flow would provide improvements for navigation and decrease costly dredging on the lower river; the massive reservoir would hold back hazardous floodwaters and make expenditures on dikes obsolete; little timber would be flooded, but the reservoir would provide water transportation to open new areas for forestry development; some land would be lost, yet greater areas would be converted to productive agriculture through the provision of irrigation and cheap power for pumps. Admittedly, fisheries "would suffer a great loss."[17] But technology could improve that problem, and in any event the comparative values were incomparable: Power, the dam's greatest product, would reach 3 million hp at the Moran dam alone.

The question that Warren did not address in his talk was how this power would be consumed. He had various uses in mind – aluminum smelting, iron foundries, irrigation pumps, and electrical heating – but provided no sense of BC's actual demand for power or the institutional complexities of its sale. He had the booster's faith: If power were provided, consumers would come. And, to judge from the experience of the U.S. Pacific Northwest, which he cited, this assumption had some basis. Electrical consumption in the Pacific Northwest states soared above that of BC during and after the war, driven by cheap power from federally subsidized dams. Why could promotional power rates not work a similar magic in BC and create a great industrial development in the process?

Although Warren's optimism knew few bounds in this discussion, his assumption of sharply rising demand paralleled forecasts of BC's power scene conducted by major institutions, governments, and private agencies in the 1950s.[18] One Department of Trade and

17. Warren, "National and International," p. 260.
18. I have depended on a synthetic treatment of these studies produced by federal civil servants in 1959; the report usefully summarizes the key estimates that I cite: BCARS, GR 1427, BC Water Rights Branch, Box 6, File 364, G. R. Knight and W. R. D. Sewell, "Evaluation of Forecasts of Electric Power Requirements in British Columbia,"

Commerce study found that, between 1945 and 1955, BC's actual increase in electrical-energy consumption more than doubled from 3.4 billion kWh in 1945 to 8.2 billion kWh a decade later. Most of this growth occurred in the 1950s: 3.5 billion kWh were added between 1950 and 1955 alone. Although part of a general North American expansion, BC's annual rate of growth for electrical consumption in the first half of the 1950s ran at 12%, or around 4% greater than the national average. How demand would increase in the future was a complicated guessing game, conditioned by unknown sources of supply (and therefore possible production and transmission costs) and the various institutional, industrial, and locational factors that segmented and conditioned BC's power markets. The Department of Trade and Commerce estimated the need for a maximum of an extra 35.6 billion kWh by 1975; the Crippen Wright Engineering firm (on contract to the provincial government in cooperation with a number of federal departments) estimated a maximum of 50.2 billion kWh for the same period.[19] Because industrial consumption accounted for over 70% of BC's total consumption in these years, differing estimates of industrial growth could swing the maximum figures in a variety of directions.[20] Yet, even if the actual figure of growth fell closer to the minimum estimates of these studies (in the range of 32 billion kWh), the growth of electrical consumption after the war gave credence to boosters like Warren. Different estimates suggested more than a tripling of electrical consumption by 1975. Where would the power come from?

The sense of inevitability surrounding a Fraser River development was also a function of decreasing alternatives. In the metropolitan regions of the province, the expansion programs of the late 1940s on Vancouver Island and at Bridge River on the mainland used up the best available sites for development within range of affordable transmission. Although these and other sites

prepared for the Inter-Governmental Technical Committee on British Columbia Power Problems, April, 1959.

19. Department of Trade and Commerce, *Electric Power Demand and Supply*; Crippen Wright, *Electric Power Requirements*.

20. Crippen Wright, *Electric Power Requirements*, Section VII, p. 3: "Industrial usage accounted for 77% of the total in 1955, and it is estimated to account for 71% of the total in 1975."

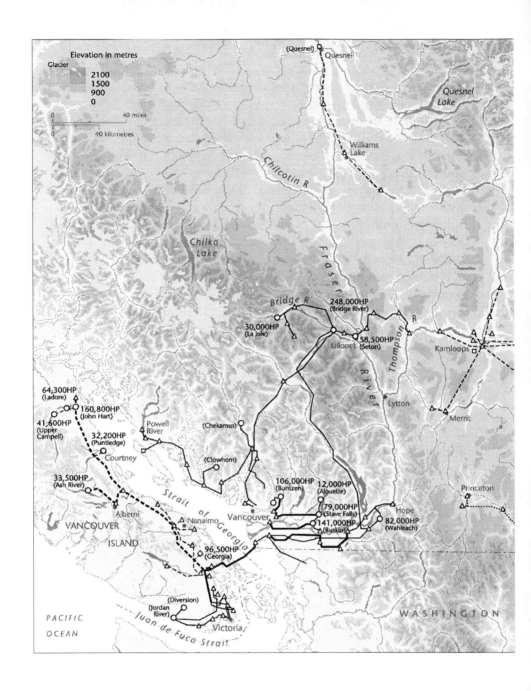

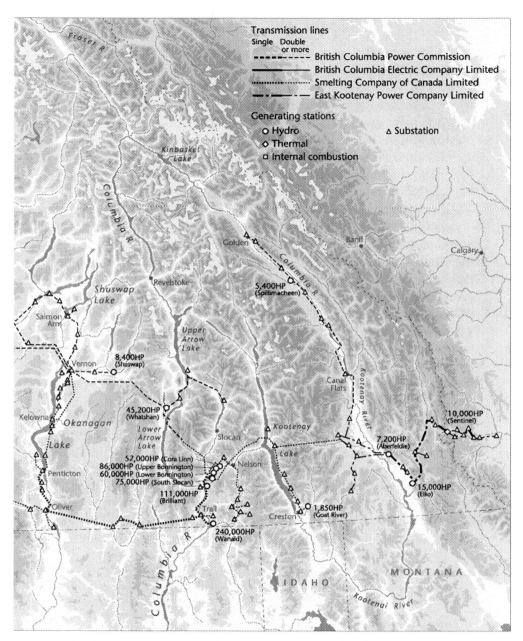

Map 8. Transmission networks in BC, 1958. Notice that the major transmission lines serving Vancouver and the Lower Mainland did not connect with lines and power sources to the east and north at this time.

(*Source:* Fraser River Board, *Preliminary Report on Flood Control.*)

191

would be expanded somewhat with generation upgrades, a new and large block of power appeared necessary by the mid-1950s, both to the private BC Electric and the provincial power commission.[21] The lower mainland market in particular seemed on the verge of major electrical demand growth according to various estimates. Studies by BC Electric beginning in the early 1950s on the engineering and economic aspects of long-distance transmission with advanced high-voltage transmission technologies suggested that capital costs and energy losses over distance restricted the range of affordable transmission to a 200–300-mile radius from Vancouver. These calculations operated on the assumptions of construction and maintenance costs, load demands, and transmission costs (based on existing high-voltage systems in other European and American jurisdictions). Power sites on the Columbia and Peace Rivers were, respectively, 400 and 600 miles distant. In the late 1950s, these rivers lay far beyond existing transmission lines to the coastal market (see Map 8). A dam at Moran, however, would be in the range of 166–206 miles, depending on the transmission route.[22] "The Fraser," stated BC Electric Vice President W. C. Mainwaring to shareholders in 1956, "is the natural next source of hydro for the Lower Mainland and Southern Vancouver Island."[23]

Warren's speech about the future of the Fraser River in 1952 resembled his wartime interventions in BC hydro debates: He wanted to spur public policy and raise the interest of private industry, but to play no direct role. That changed in the mid-1950s. The professor turned promoter. Starting around 1955, Warren acted as one of the directors of the Moran Development Corporation. The Canadian directors included Russell Potter, an engineer and formerly executive assistant to the Fraser Basin Board, who had helped Warren

21. Premier Bennett was kept abreast of the BC Power Commission's fears of a supply shortfall: Simon Fraser University (hereafter SFU) Archives, W. A. C. Bennett Papers, Box 8, File 4-6, T. H. Crosby, chairman of the BC Power Commission to premier, January 12, 1956; "Notes on BC Power Commission on Vancouver Island for the Honourable Premier," May 16, 1956.
22. Steede, "The Long Distance Transmission of Energy." On the problems of long–distance transmission, see also Walker, "The Cost of Electrical Energy Generation and Transmission."
23. SFU Archives, W. A. C. Bennett Papers, Box 8, File 3, "Address to Shareholders of British Columbia Power Corporation, Limited, March 29, 1956" ('Mainwaring, BCE' written at top).

with the technical detail of his first Moran paper, as well as Harry
Swinton, a Vancouver lawyer. The financial backers were Americans, principally Hans Eggerrs, formerly an executive with the
Continental Can Corporation, and Alfred Vang, an inventor of dubious ideas, but spectacular promotions.[24] The corporation aimed
to develop a dam at the Moran site and produce multiple-use developments along the lines that Warren had earlier outlined. Of
all of the individuals involved, Warren took the greatest public
role, speaking, it seemed, to any group that would listen and publishing numerous articles in the engineering, mining, and business
press.[25]

Promotion, however, did not equal possession. Although the
provincial government granted the Moran Development Corporation rights to explore the site for drilling and engineering studies
in 1955, no reservation was granted, as had been the earlier practice
in the Alcan case. Within months, BC Electric, acting to displace its
new competitors, also received provincial permission to examine
the site, thus removing whatever priority the Moran interests had
once hoped to gain.[26] The attempts to secure financial backing appeared equally illusory. In planning meetings with provincial civil
servants, Warren and his Canadian colleagues gestured about forthcoming financial backing, if only some agreement could be resolved

24. Metcalfe, *A Man of Some Importance*, p. 197.
25. Several articles are contained in Warren's papers held at UBC: "The Power Potential of
    the Fraser River," *BC Professional Engineer* 3(4) (1952): 19–23; "Power, Population
    and Politics," *BC Professional Engineer* 3(10) (1952): 25–32 and another paper under
    same title in *BC Professional Engineer* 3(11) (1952): 22–8; "Energy for Everyman,"
    BC Professional Engineer 4 (1953): 19–23; "Hydroelectric Potentialities of the Upper
    Fraser," *Western Miner and Oil Review* 29(3) (1956): 32–7, and reprinted in *BC Pro-
    fessional Engineer* 7(7) (1956): 16–24; "Background for Crises," *Western Miner and
    Oil Review* (1957); "Moran Dam Holds Key," *Western Business and Industry* 32(8)
    (1958): 56–60; "Moran Dam," *Northwest Digest* 12(4) (1956): 9, 30–5, reprinted in
    Canadian Mining Journal 80(3) (1959): 63–8.
26. BC Water Management Branch, Ministry of the Environment, Lands and Parks, Micro-
    films of the Department of Lands 'O' File correspondence, File 0188688, Roy Williston,
    minister of lands and forests to Tom Ingledow, vice president, BC Electric, July 16 1956
    (copy); Williston to A. Hans Swinton, Moran Development Corporation, July 25, 1956
    (copy). BC Electric management, nevertheless, feared that priority might be granted to
    the upstart concern. In a 1955 letter to Premier Bennett, W. C. Mainwaring responded
    to what he believed was a radio announcement suggesting that permission to proceed
    had been granted; he hoped the site would be reserved for the needs of the Lower Main-
    land: SFU Archives, W. A. C. Bennett Papers, Box 6, File 8, Mainwaring to Bennett,
    August 9, 1955.

as to dam design and the fish–power problem.[27] The chief American backer, Hans Eggerrs, provided little help in this respect. He had recently been dismissed as an executive by the Continental Can Corporation for extravagant research expenditures, paid out to prove (unsuccessfully) the heterodox metallurgical ideas of Alfred Vang.[28] These reputed "American financial backers" could provide only the status and mystery of outside capital, without any of the financial clout. The best that the Moran Development Corporation could have hoped for in 1955 was to be bought out by BC Electric, and this may well have been the intention. This, at any rate, would have suited Warren's purposes: Moran was the goal, the corporation a vehicle. If the three pillars needed to hold up the Moran plan were promotion, politics, and finance, only the first appeared steady in 1955.

The promotion, nevertheless, continued. The Moran idea gathered a disparate collection of supporters and worried competitors in the 1950s.[29] Before the demise of the coalition government in 1952, Minister of Public Works E. H. Carson committed himself to the project.[30] In 1954, writers in *both* the Labour Progressive (Communist) Party paper and the *Victoria Colonist* wrote approvingly of Warren's plans. In the same year, Social Credit Minister of Lands and Forests Robert E. Sommers, who would later fall into disgrace for taking bribes from forestry companies, made a speech extolling the idea of Fraser dams and looked to the region above Lytton for future development.[31] BC Electric and the federal–provincial Fraser Basin Board rushed to catch up with the Moran Development Corporation, pursuing a series of feasibility

27. BCARS, GR 1118, BC Marine Resources Branch, Box 3, File 1, "Notes on Meeting with Moran Power Development Ltd. held in the offices of the Chief Supervisor of Fisheries on May 24, 1956 at 11:00 am."
28. Metcalfe, *A Man of Some Importance*, p. 197.
29. Including a substantial MA thesis, in part funded by the Corporation: Hardwick, "The Effect of the Moran Dam."
30. BCARS, GR 1378, BC Commercial Fisheries Branch, Box 5, File 5, *Cariboo Observer*, June 7, 1951; *Williams Lake Tribune*, June 7, 1951.
31. BCER CF, Hal Griffin, "Harness the Fraser? A People's Government Would Do It" (nd. *Pacific Tribune?*). The article mentions Warren's plan; Alf Dewhurst, "Harness the Fraser," *Pacific Tribune*, June 4, 1954; "Damming the Fraser," *Columbian*, March 4, 1954; T. A. Myers, "Hydro Potential Hardly Touched," *Victoria Colonist*, March 3, 1954. The last two articles mention Sommers's views.

studies.[32] The IPSFC focused on the consequences of the project and began to compile documents assessing impacts.[33] The Moran concept appeared in the press of BC's interior, where local notables wrote glowing copy celebrating a future of big projects and local development, in metropolitan dailies, where the trade-offs of power and fish commanded attention, and even in the international press, where the *New York Times* announced the project to the world.[34] From its early beginnings at the Natural Resources Conference, the Moran idea had taken on promotional, if not material, form by the mid-1950s.

The Columbia River Question

If the Fraser figured in public discourse as the river of inevitable development, then the Columbia did so as the river of perpetual delay. Although the American portion of the river began to be developed in the early 1930s, no main-stem development had occurred in the Canadian upper basin. Since the mid-1940s Canada and the United States had cooperated through the IJC in the investigation of storage possibilities in Canada. Founded in 1909, the IJC was a unique binational institution comprising six members, three from each country, with the primary task of studying and advising governments on transboundary water issues. It embodied the ideals of progressive conservation, granted significant authority to experts, and aimed for binational cooperation and conservation.[35] The Columbia River posed a significant challenge to the IJC and provoked considerable internal and external debate over how two neighboring nations should coordinate river uses. American IJC

32. NAC, RG 89, Box 674, File 2516, "Preliminary Study of Moran Canyon Project on Fraser River," prepared for Fraser River Board by Crippen Wright Engineering Ltd., Engineering Consultants, Vancouver, BC, April, 1956.
33. IPSFC, *Annual Report* (1956), p. 28.
34. BCARS, GR 1378, Commercial Fisheries Branch, Box 5, File 5, editorial, *Cariboo Observer*, June 7, 1951; editorial, *Williams Lake Tribune*, May 31, 1951; editorial carried from *Prince George Citizen* in *Cariboo Observer*, June 14, 1951; BCER CF, "US Northwest to Gain from Projects to Develop British Columbia Water Power," *New York Times*, January 6, 1954.
35. Brown and Cook, *Canada 1896–1921*, pp. 174–7.

negotiators pressed for a fully integrated development program that would treat river regulation as a whole and optimize downstream uses. Canadian negotiators realized that to do so would diminish other potential upstream uses; coordinated development would require compensation. By 1954, disputes within the IJC over the appropriate means to compensate Canada for turning the upper basin into American storage ruled out a rapid development schedule. Outside the IJC, five different groups proposed upper-basin projects – with no concern for basinwide coordination – and failed to receive the necessary binational support. It was in this atmosphere of stalled development and international complication that a new Columbia plan emerged (see Map 7).

What if the Columbia were diverted into the Fraser? This was the question put in late 1954 by General Andrew McNaughton, Canadian chair of the IJC. Unlike Professor Warren, who was influential in BC's resource economy but held no discernible political power, McNaughton approached promotion from a position of authority with important allies. The chair of the Canadian Section of the IJC since 1950, McNaughton held influence through his office; his reputation earned in the war and as the head of the National Research Council in the 1930s gave him a national profile. In his annual reports to Parliament, McNaughton cut such a distinguished profile that numerous commentators assumed that he not only represented but also established Canadian policy on the Columbia, and the assumption held some truth.[36] Before the mid-1950s the federal government (to the province's annoyance) remained aloof from the Columbia negotiations, leaving McNaughton considerable room to make decisions. When in 1954, the BC government sought to press forward Columbia development with the Kaiser project, McNaughton's appeal to the federal government that the scheme contradicted the national interest held sway. On his advice, the federal government pursued the cancellation action.[37] It was because of this elevated authority, well placed and politically connected, that McNaughton's proposal for the Columbia in 1954 did not face the immediate death of so many other postwar Columbia

36. On the importance of McNaughton's reputation to the conduct of his work, see Swainson, *Conflict Over the Columbia*, pp. 50, 53–4, 64.
37. Swainson, *Conflict Over the Columbia*, p. 59.

schemes. In lesser hands, with weaker political allies, it is difficult to imagine that a plan of such magnitude, indeed hubris, would have appeared so possible.

Whereas the pressure for development on the Fraser grew out of BC's soaring electrical demand and forecasts of continuing growth, on the Columbia the pressure came ultimately from American interests. The IJC's investigations began in 1944 at the request of the U.S. government with a view to future power needs and transnational coordination. After the flood of 1948 and the continued rise in power demand in the U.S. Pacific Northwest after the war, the lure of the Columbia only increased in American eyes. Whereas no Canadian developers proposed projects on the Canadian Columbia in the first decade of IJC studies, five American groups did.[38] There were, of course, considerable rewards to be reaped by Canada in a coordinated development program. This is what inspired the active involvement of the provincial government in Columbia negotiations after 1950 and its support of the Kaiser proposal in 1954. The provincial position, however, was premised on the understanding that Columbia development would proceed only with American investment and involvement.

General McNaughton's proposal envisioned an entirely different course of action. His idea was to keep Canadian water for Canadian power development and dispense with the complications of international coordination. In late 1954, he made his views known in confidence to a select group of politicians and power company executives. By capturing the upper Columbia at Mica Creek, he explained, storage could be created. Instead of releasing this water for downstream purposes, it could be diverted by means of a pipeline through the Eagle Pass into Shuswap Lake and thence into the Thompson River, the Fraser's largest tributary. Skimming the Columbia's high flows during the spring, the diversion would place this excess into the Fraser during the low-flow season, evening out the Fraser's fluctuations and making for a steady power stream. Low-level dams on the Thompson and Fraser, perhaps ten in all, would catch this extra flow and provide BC with its future energy

38. A Canadian project sponsored by Consolidated Mining Co. on the Pend Oreille River, however, did have an impact on IJC discussions on the Columbia.

needs. Because the dams would be small in scale, below 100 ft in height, McNaughton believed no damage would come to the salmon fishery. By a confluence of continental flows, the general argued, continentalism in water planning could be abandoned and Canada's national ambitions realized.[39]

One of the general's key supporters in this revised plan for Columbia development was Jean Lesage, minister of northern affairs and national resources.[40] Before the plan became public, Lesage committed his department to fund surveys in the Columbia Basin, examining diversion points and other feasibility aspects in the amount of $200,000. Although Lesage appeared to find the plan intriguing and promising, he no doubt understood the political payoffs that would result even from the investigation of such a plan. Throughout negotiations, Americans had sought to point out that Canada's alternatives to coordinated development appeared minimal and that American compensation payments to Canada should reflect this point. By proposing to divert the Columbia, Canada's argument for alternatives gained substance and so too its calls for compensation. The other attractive aspect of the diversion proposal from a federal perspective was that it offered British Columbians an impressive alternative to the provincially backed Kaiser project. Whereas the provincial plan would afford development rights to an American interest at a low cost, the McNaughton plan would keep Columbia power in Canada for BC's purposes. At a time when the province was bemoaning the federal intervention into the Kaiser proposal – Premier Bennett called it "a cheap political trick" – the new plan could show that the federal government was concerned not with delaying development, as the province contended, but with making Columbia development as propitious as possible for the province and the nation.[41]

---

39. SFU Archives, W. A. C. Bennett Papers, Box 5, File 4, A. G. L. McNaughton, chairman, Canadian Section, IJC to the secretary of the treasury board, November 2, 1954 (marked confidential) (copy). Although McNaughton's initial statement of the plan made no reference to it, he would later include a diversion from the Kootenay River into the Columbia as another aspect.
40. Swainson, *Conflict Over the Columbia*, p 61; BCER CF, "Fraser Harnessing for Hydro Probed," *Vancouver Sun*, December 20, 1954.
41. BCER CF, "Bennett Blasts Fisheries Minister," *Victoria Times*, November 24, 1954.

McNaughton linked his plan not only with federal interests, but also with the province's major utility, BC Electric. Whereas Harry Warren's Moran plan sought to compete with the established utilities, McNaughton sought to involve them. In November 1954, McNaughton met with Dal Grauer and Tom Ingledow, respectively the president and vice president of BC Electric, and asked them to consider the diversion idea. At the time, BC Electric was busily trying to examine development prospects for power growth over the next two decades. They considered sites on the Fraser below Lytton, as well as an extension at the Bridge River facility and power-sharing agreements with rival utilities. The diversion concept and its promise of major power potential caught their immediate attention. Before the end of the meeting they had secured McNaughton's agreement to allow their company to carry out confidential survey work on contract and made it clear that they wished to cooperate with the general in pursuing the plan. The project, they explained, could deliver BC's metropolitan electrical needs in both the short and the long term. The initial development at Mica Creek would be used to meet the current rise in demand over the next seven years. The Thompson–Fraser dams to follow would satisfy the company's needs for the next two decades. The two-stage nature of the project also promised to allow additional time to solve the fish–power problem on the Fraser. The unstated benefit to be gained was that BC Electric would have, by virtue of its early involvement, a de facto privilege and priority over other power concerns.[42]

McNaughton's ambitions, nevertheless, ran headlong into provincial plans. The provincial government remained disappointed by the actions of McNaughton and the federal government in the Kaiser affair and saw the diversion as yet another form of federal interference. Instead of embracing McNaughton's option, even

---

42. SFU Archives, W. A. C. Bennett Papers, Box 5, File 4, "Memorandum Re Meeting with Representatives of British Columbia Electric Company," November 1, 1954; Dal Grauer spelled out the company's interest in a follow–up letter: NAC, RG 23, Vol. 1229, File 726-11-10[1], A. E. Grauer, president of BC Electric, to McNaughton, December 19, 1954. Bennett, however, tried to make it clear to BC Electric that they should make no assumptions about priority, as BC Electric Vice President Tom Ingledow informed Jean Lesage: Box 9, File 4, Ingledow to Jean Lesage, June 5, 1956 (copy to Bennett).

as a short-term political maneuver to enhance Canada's bargaining position, the provincial government gave it a wide berth and cast aspersions to the press through the medium of anonymous leaks. The diversion was described as "impractical, unnecessary and too costly" and "[c]ompletely fantastic, pure sheer nonsense."[43] These anonymous provincial critics pointed to the potential losses to the fishing industry and said that the move would only stall development. When Attorney General Robert Bonner testified before a House of Commons committee concerning the Kaiser project cancellation in 1954, he let it be known that the BC government held "only academic interest" in the diversion proposal.[44]

Another voice of opposition arose from within the federal government, upsetting the easy assumption of federal support and provincial opposition to McNaughton's plan. Despite its attractiveness to Canada's negotiating position and the promise it held for BC development, BC MP and federal Minister of Fisheries James Sinclair soon registered his department's unease at the proposal.[45] The IPSFC, who informed McNaughton confidentially of the enormous risks to the salmon fisheries, seconded him in these concerns.[46] Sinclair was in a difficult position publicly; he had been the BC Liberal to announce the federal opposition to the province's Kaiser proposal; now he would have to choose between supporting the federal solution to the Columbia problem and advocating the fisheries concerns in keeping with his ministerial obligations. Initially, he managed a delicate balance, supporting investigations presumably to enhance Canada's negotiating hand, but establishing the potential losses to the fisheries through interdepartmental committees and prevailing upon fishing interests in BC to prepare for a political battle. It was a balance that could last only so long as the diversion remained a concept and not a pressing reality.

43. BCER CF, "The Wasted Water," *Victoria Times*, December 23, 1954; "Ottawa Plan for Fraser River Power Derided by Provincial Authorities," *Victoria Colonist*, December 19, 1954.
44. Quoted in Swainson, *Conflict Over the Columbia*, p. 62.   45. Ibid., p. 68.
46. NAC, RG 23, Box 1229, File 726-11-10, part 1, "Confidential Statement by the Chairman of the IPSFC, on Behalf of that Commission, to General A. G. L. McNaughton, Chairman, Canadian Section, IJC, and to the Governments of Canada and the United States of America," June 3, 1955.

From an American perspective, the diversion proposal inserted an entirely different view of Columbia development than had existed previously and threatened a variety of national interests. If the diversions went ahead, not only would coordination vanish and potential harm come to downstream projects, but America's fishing interests and involvement in Fraser River restoration would also be affected. Senator Richard Neuberger of Oregon brought out these possibilities publicly in the United States in 1955 after touring BC on a fact-finding mission at the request of Senator James E. Murray of Montana, the chair of the U.S. Senate's Interior and Insular Affairs Committee. Neuberger had long taken an interest in Columbia affairs; in his former career as a journalist he had penned some of the most enduring stories of the river, encapsulated in his book, *Our Promised Land*, a tract of New Deal dreams, with the Columbia playing the role of regional savior.[47] He found the diversion plan disturbing and all too possible. "That this is not merely an empty gesture," he later informed members of the Senate committee, "is verified by the fact that Gen. A. G. L. McNaughton – the illustrious soldier–general who is Chairman of the Canadian Section of the International Joint Commission – thoroughly believes in the engineering feasibility of the diversions, and that the Canadian Parliament last year voted a very substantial appropriation to follow through on the engineering studies of the diversions and to try to establish their economic feasibility."[48] Further, unlike other American commentators who questioned the legal basis of diverting an international river, Neuberger argued that Canada was within its rights: the International Waterways Treaty of 1909, the legal basis of the IJC, set out the priority of the upstream nation in water development.[49] Unless the United States could convince Canada of the benefits of coordination, and compensate Canada appropriately, the future of the Pacific Northwest states would be hindered:

47. Neuberger, *Our Promised Land*, introduction by David L. Nicandri.
48. "Joint Hearings Before the Committee on Interior and Insular Affairs and a Special Subcommittee of the Committee on Foreign Relations, United States Senate, Eighty-Fourth Congress, Second Session, March 22, 26, 28 and May 23, 1956" (Washington, D.C.: Government Printing Office, 1956), p. 4.
49. Neuberger expressed his views on the legal question of diversion in a 1957 *Harper's* article reported in BCER CF, "BC Neighbors Alarmed in Border Power Fight," *Vancouver Sun*, December 11, 1957. On the general legal discussion on diversion, see Swainson, *Conflict Over the Columbia*, pp. 65–7.

"It means," Neuberger stressed, "the difference between economic progress or stagnation."[50] Interestingly, Neuberger laid most of the blame for this situation with the bargaining stance of the U.S. negotiators within the IJC, particularly American Section Chair Leonard Jordan, formerly the governor of Idaho. He accepted the Canadian claim for substantial compensation, on the other hand, as eminently reasonable. Neuberger's forthright explication of the issues raised the profile of the problem and inserted the strong regional concerns of the Pacific Northwest in the conduct of the U.S. position within the IJC.[51] The diversion plan had at last brought the matter of downstream benefits before American legislators.

Harry Warren's Moran concept and General McNaughton's diversion scheme were two of the most prominent plans to dam the Fraser aired in the mid-1950s. But they were exemplars of a movement rather than its only driving forces. In these years, BC Electric investigated the Moran site, the Columbia diversion dams, and other sites on the Fraser's main stem. At Seton Creek, the company extended its earlier Bridge River project and dammed a pink salmon run in the process.[52] The BCPC examined the possibility of damming Taseko and Chilko Lakes – formerly the site of intense fisheries opposition during the Alcan surveys – and applied for water rights at the location.[53] As discussions on the Columbia stalled, the BCPC also investigated the Clearwater River, a Thompson tributary.[54] Throughout the upper basin, the Fraser Basin Board conducted studies for multiple-purpose dams to stem future flood threats and generate power. The river, in short, was under active scrutiny by engineers and power concerns, and the oft-stated

50. Ibid., p. 6.
51. Swettenham, *McNaughton*, p. 257. Swettenham claims that the reaction to Neuberger's claims in the U.S. Senate was "near panic." This strikes me as a clear overstatement, given the record of the hearings. Swettenham provides no other evidence to support his claim.
52. UBC Special Collections and Archives, Fisheries Association of BC Papers, Box 45, File 45-11, C. H. Clay, "The Seton Creek Project."
53. NAC, RG 23, Box 842, File 719-9-92[1], A. W. Lash, consulting engineer, BC Power Commission to A. J. Whitmore, October 23, 1956; Loyd Royal, IPSFC, to Whitmore, January 11, 1957; J. M. Buchanan, chairman, Fisheries Association of BC, to Whitmore, February 1, 1957; "Notes on Meeting with BC Power Commission to the Fisheries Problems Associated with the Chilko–Taseko Project," March 4, 1957. IPSFC, *Annual Report* (1956); Williston and Keller, *Forests, Power and Policy*, p. 181.
54. Wedley, "The Wenner-Gren and Peace River Power Development Programs," p. 526; BCER CF, *Vancouver Sun*, February 4, 1957.

warning or threat that the river would soon be dammed appeared undeniable. What remained for British Columbians to determine was the shape and scope of such development. Would there be fish or power? Could there be both?

Boundaries of Debate

The fish vs. power debate of the 1950s proved to be an expansive discussion. The narrow problem of articulating conflicting resource interests in particular instances turned into a debate over the future of society and its relations with nature. It invited British Columbians to consider the merits of development and growth, as well as their costs; to determine how favoring fish or dams would shape BC as a region both internally and in relation to external influences; and to ask whether alternatives existed – coordinated development, alternative energy sources, or scientific panaceas. British Columbians shared with Harry Warren the belief that damming the Fraser would forever transform the river and themselves. In academic conferences, the legislature, in kitchens, high school debates, and letters to the editor, British Columbians anxiously considered this transformation. In seeking to imagine and shape the future, they revealed much about their present condition and predicament.

Although the fish vs. power debate involved more than industrial interests, it was also the case that these interests dominated debate. On the fisheries side, the Fisheries Association of BC and the UFAWU played the most prominent roles, and politicians and officials attached to the federal Department of Fisheries and the IPSFC worked to organize the industry and combat indiscriminate water-development policies within government. Together these groups and others formed a Fisheries Protection and Development Committee in 1956 under the auspices of the Department of Fisheries to coordinate their actions.[55] This committee provided a forum for

55. This group changed its name over time to the Fisheries Development Council. The origins of the group are described in the preface to "Summaries of Research on the Fish-Power Problem and Related Work by Fisheries Agencies in British Columbia" (Vancouver: Department of Fisheries, Revised December 1961)" contained in BCARS, GR 442, BC Energy Board, Box 52.

the dissemination of information and the coordination of political tactics among different groups from the Native Brotherhood to the UFAWU, to the IPSFC, and to sports fishers. In the past, these groups had rarely cooperated, and several, such as the Fisheries Association of BC and the UFAWU, had contentious relations. That these groups managed to join forces on the fish vs. power issue suggests how seriously they judged the threat.

The power side of the debate, on the other hand, was less well organized, in part because its various elements were competitive with one another. Corporations weighed heavily in the discussions: BC Electric, the BCPC, and the Moran Development Corporation made parallel, but distinct interventions. At the government level, the federal Department of Northern Affairs and National Resources, the IJC, and the provincial Ministry of Lands and Forests generally favored power positions, but played no role like that of the Department of Fisheries to organize a power bloc. The only coordinated lobby group to appear on the power side was the Fraser River Multiple Use Committee, started in Vancouver in 1958 to consider the many different demands on the Fraser and foster cooperation. It was closely associated with the Moran proposal, however, and this limited its appeal. The only group that could be said to bridge the divide between fish and power interests – and then only barely – was the Fraser Basin Board; it contained both water development and fisheries representation in planning flood-control measures.

Despite the influence of the fish and power groups on the broader discussion, it is important to recognize the extent to which the fish vs. power issue transcended traditional party lines and fractured political interests. The neat division apparent in the title of the debate, and seized on by participants and the media alike, masked a range of possible positions on these issues and much public confusion about the choices. One newspaper reporter observed in 1957 that the fish vs. power fight was "the strangest industrial dispute in BC history"; it was so unpredictable in its twists and turns and involved so many unexpected entrants that a program would be necessary to follow the confrontation.[56] At the federal and provincial levels no governing parties could demonstrate consensus on the

56. BCER CF, "First Round Over in Power Dispute," *Vancouver Sun*, May 28, 1957.

issue, nor could parties in opposition. Federally, the Department of Northern Affairs and National Resources funded surveys for Fraser dams, while the minister of fisheries organized the fishing industry to protest them. Provincially, prominent cabinet ministers, such as Minister of Lands and Forests Ray Williston, intimated support for Fraser dams, while the premier studiously avoided making strong commitments.[57] All parties, one journalist argued, "[are] split on power for the Fraser."[58] This level of political confusion only stoked the fires of possibility in the broader discussion: Nothing was decided, the political discussion was unresolved, and the problems forcing discussion forward – looming power demand and Columbia River negotiations – only heightened the tension.

For all of the ways that the fish vs. power debate connected to the contemporary BC context and the dynamics of its political culture, it also contained revealing absences. In over a decade of newspaper coverage, academic commentary, and political discussion, the potential effects of Fraser River dam development on native fisheries received little attention. Although one might have expected the precedent of the Cheslatta flooding case during the Alcan project to provoke discussion about the consequences of dam development on native peoples, this was not the case. Fisheries groups did seek to include the Native Brotherhood in their political organizing and deliberations, but the Brotherhood's membership was based on the coast and did not represent interior groups along the river. On several occasions, the Brotherhood did attempt to assert the importance of the native food fishery. But they did so without much external recognition. The fish vs. power debate operated with a remarkable detachment from native perspectives.

Debating Meanings

Power promoters promised many things of Fraser River dams, but perhaps the most intangible reward offered, and also the most often cited, was progress. Since the earliest days of spectacular lighting in

---

57. BCER CF, "Fraser Dams Being Studied," *Vancouver Sun*, February 26, 1958; "Dams and Fish May Mix," *Victoria Colonist*, November 9, 1958; "Fraser Dams a Must," *Chilliwack Progress*, June 16, 1961.
58. BCER CF, "Parties Split on Power for the Fraser," *Province*, December 18, 1957.

the late nineteenth century, electricity evoked a futuristic language of new beginnings in North America. "In the 1930s," American historian David Nye writes, "electricity was still a new technology that suggested radical change. Most could still recall the pre-electric world, and advertising abetted historical memory with images suggesting how the electric present differed from the past, and which predicted even greater, immanent transformations."[59] BC in 1950 held to the idea of electricity's newness and promise. After the brownouts and electrical restrictions of the war years, the creation of the BCPC and the start of BC Electric's expansion projects in the late 1940s, British Columbians waited expectantly for progress to arrive. Newspapers greeted hydro development as a "source of progress," "of future strength," a "step forward," the unlocker of "future's door," the "modern means to industrialization," "and a modern and efficient prime mover."[60] In 1954, the *Province* imagined a future made possible by the Columbia–Fraser diversion, in which a doubling of "the industry and population of BC" would occur "in 15 to 20 years." "It would fulfill the dream of a new industrial empire in this province."[61] Not coincidentally, the BCPC's advertising slogan was "power means progress."

The ubiquitous association of these two words was also conditioned by the particular contexts of time and place. The assertion of power as progress in BC of the 1950s contained a sometimes explicit, often times implicit, pairing: fish, the obstruction to power, represented the past, stagnation. In 1957, for example, Diana Davidson, a North Vancouver high school student, wrote to the editor of the *Province*, instructing her fellow citizens to "See that you know the facts of the Fraser River power issue and then support power and progress." This confident advice sprung from Davidson's recent triumph over her suburban West Vancouver rivals in a "Fish vs. Power" debate sponsored by federal Minister of

---

59. Nye, *Electrifying America*, pp. 339–40; see also Nelles, *The Politics of Development*, Chapter 6.
60. BCER CF, "Whatashan: Source of Progress, Prosperity," *Vernon News*, June 28, 1951; "Rural Electrification Takes a Step Forward," *Province*, June 28, 1951; "Great Day Dawning," *Kamloops Sentinel*, January 17, 1949; "Electricity is 'Open Sesame' to Okanagan Valley Growth," *Vancouver Sun*, March 28, 1950; "Interlocking Pattern of Power Transforms Island's Economy," *Victoria Colonist*, October 26, 1949.
61. BCER CF, "Power for an Empire or for Peanuts?" *Province*, December 30, 1954.

Fisheries and local MP James Sinclair. In her advocacy of the power position, Davidson drew an implicit comparison between power as progress and fish as past.[62] This meaning was drawn more explicitly in Premier Johnson's 1950 statement that British Columbians must choose between power development and *"remaining as they are* protecting the fisheries."[63] The pairing also opened a rhetorical space for critics of dam development to be represented as opponents of progress. "There is a growing suspicion," wrote the editor of the *Cariboo Observer*, "that hydro is just plain unpopular with certain interests that are willing to stand in the way of progress."[64] To believers in the inevitability of power and progress, this opposition, of course, stood against the unstoppable. "Irresistible forces of unfolding history," stated General Victor Odlum, a veteran and former Canadian diplomat in 1954, "will sweep us on anyway to a key position in the great Pacific civilization of the future." A key force, he said, would be the Fraser: "probably the greatest single potential power producer on the continent."[65] Thus the statement of power as progress contained a barbed edge: Its optimism discounted the fisheries claim for legitimacy.

Although fish could be dismissed as past, however, they could also be celebrated as an enduring connection with tradition, heritage, and nature. By the 1950s, salmon had become an evocative symbol of regional identity. BC's most famous author in this period, Roderick Haig-Brown, had made a career out of writing about the pleasures of fishing and the life histories of different fish for a leisured middle-class readership.[66] In *The Fraser*, journalist Bruce Hutchison titled his chapter on the fisheries, "the first inhabitants." Rhetorically eliding a native claim to this status, Hutchison constructed salmon as subjects of history, "our" connection to an organic, primordial BC past.[67] In his criticisms of indiscriminate power development in the late 1940s and 1950s, IPSFC chairman

---

62. BCER CF, letter to editor by Diana Davidson, *Province*, May 23, 1957.
63. BCER CF, "Fisheries or Power?" *Vancouver News Herald*, May 19, 1950. Emphasis added.
64. BCER CF, "Hydro Seems Unpopular," *Cariboo Observer*, nd.
65. BCER CF, "BC Too Timid to Pursue Destiny," *Province*, February 19, 1954.
66. Keeling and McDonald, "The Profligate Province"; Metcalfe, *A Man of Some Importance*.
67. Hutchison, *The Fraser*, Chapter 19.

Tom Reid railed against the "steal[ing] of our heritage" on the Fraser River.[68] At the end of the decade, during the 1958 BC centenary celebrations, the provincial fishing industry collaborated to produce "Salute to the Sockeye" festivals, events that attempted to remind British Columbians of the historic qualities of the fisheries and the importance of salmon in the past and the present.[69]

Fisheries propaganda played on an anthropomorphized representational strategy. A pamphlet produced by the UFAWU in the mid-1950s, for example, invited British Columbians to protect salmon as the victims of progress: A cartoon fish on the pamphlet's cover nervously eyed a dam in its path as it sprang from the river, crying "Well I'll be dammed!" The image was meant to amuse but also encouraged readers to sympathize with the salmon's anxiety.[70] In an amusing send-up of the fish vs. power debate in 1958, one ironic *Vancouver Sun* reader penned a short letter, signed "Samuel Sockeye" that assumed the voice of an individual sockeye, speaking on the power issue for "all finny denizens of the Fraser."[71] Imagining the private lives of salmon provoked laughter, but it also signaled the extent to which British Columbians represented salmon as subjects worthy of sympathetic moral imagination.

The value of salmon was pressed further by the assertion of the materiality of fish as *food* against the ethereal promise of electricity as progress. In a speech to the BC Natural Resources Conference in 1951, A. H. Sager of the BC Fisheries Association stressed the renewable quality of salmon: "The fishing industry was the first industry, it was the means of livelihood for the Indians long before the white people came. It fed the people of Galilee. I think it probably fed the cavemen. And I believe . . . that the fisheries of our coast will be feeding British Columbians and Canadians 100, 200, 300 years from now, when, perhaps, hydro-electric installations have become obsolete."[72] For a society that held unpleasant

68. NAC, RG 23, Vol. 1570, File 784-3-3, part 3, Minutes of a Meeting of the IPSFC held at Vancouver, BC, December 8 and 9, 1955.
69. UBC, Fisheries Association of BC Papers, Box 23, File 23-8, Minutes of Fisheries Protection and Development Committee, April 7, 1958; Annual Report, 1957.
70. UBC, United Fisherman and Allied Worker's Union Papers, Box 137, File 137-2, pamphlet, "Well I'll Be Dammed: A Fish Story," nd.
71. BCER CF, "From a Fish," *Province*, January 28, 1958.
72. "Forum: Fish and Power," *Transactions of the Fourth British Columbia Natural Resources Conference* (Victoria, February 22, 23, 1951), p. 111.

memories of the Depression and the sacrifices of the war years, this statement of salmon's value as food had important meaning. Numerous letters to the editor in the 1950s stated that it would be wrong to destroy food in a world filled with want; the *Victoria Times* called the destruction of salmon as food a "moral crime."[73] Fisheries supporters sought to harness this attitude with the slogan, "You can't eat a kilowatt." Employed in propaganda literature by the UFAWU and by federal Minister James Sinclair, the phrase underlined Sager's point that an electrified future would never provide food.[74] As to the future, the importance of salmon as food would only increase. Facing the challenges of the cold war, the *Columbian* imagined the prospect of nuclear annihilation in 1954. "In such a disaster, stricken peoples cannot get food from broken machinery. Land yields food, but it has to be tilled. Fisheries may save countless lives."[75] To counter such connections, power promoters felt forced to explain how electricity would create more to consume, not less. Harry Warren declared in 1960 that, for every pound of salmon lost to dams, there would be two pounds of beef created.[76] In a televised appearance in 1959, Gordon Shrum, a UBC physicist and recently appointed head of the BC Energy Board, dismissed "salmon romanticism" and drew an analogy between the clearing of bison on the plains to make way for wheat and the removal of salmon on the Fraser to allow for power development.[77] The best that power promoters could say in response to the kilowatt slogan was, "You can't burn a fish."

Beyond celebrating salmon "heritage" and the moral significance of salmon as food, fisheries defenders sought to portray the industry as restored, future oriented, and growing. Thus Loyd Royal, the director of scientific investigations for the IPSFC, stated to the BC Natural Resources Conference in 1954[78]:

The 1953 run was the largest cycle run since 1912 and the catch in 1951 was the greatest in the cycle year since 1903. The value of the catch of the last three years exceeded that of the preceding three year cycles by almost

73. BCER CF, "Both Fish and Power," *Victoria Times*, December 17, 1957.
74. BCER CF, "Silly Slogan Imperils Salmon," *Vancouver Sun*, September 12, 1958.
75. BCER CF, "Damming the Fraser," *Columbian*, March 4, 1954.
76. BCER CF, "Fraser Dam Held Gain Despite Fish," *Vancouver Sun*, February 24, 1960.
77. BCER CF, "Of Buffalo and Fish," *Vancouver Sun*, November 6, 1959.
78. Royal, "The Rebirth of the Fraser Sockeye."

$18,000,000. This is a considerable sum but it is only a start toward the foreseeable goal of re-establishing the original economy of the Fraser River sockeye fishery. The once great Quesnel run is firmly re-establishing itself. The Stuart system of the far north produced over 2,500,000 sockeye in 1953 yet the total escapement in 1941, only three cycles previous, was less than 12,000 fish. The rebirth of the Fraser sockeye in dollars and sense has truly commenced.

This rebirth, fisheries scientists and officials assumed, would continue. "If currently known methods of conservation were fully applied," a provincial report stated in 1955, "the total catch of salmon could probably be doubled."[79] The actual experience did not bear out these predictions, but throughout the late 1950s, salmon numbers continued to climb. Dianne Walsh reported in the *Columbian* that 500 times the numbers of fish as the same cycle twenty years ago would return to the Fraser in 1961. "Fish," the headline claimed, are "proving their own case in [the] Fraser River power fight."[80] The growth prospects for the fisheries, argued one fisherman in 1958, should make British Columbians reconsider the relative economic values of fish and power.[81]

If a reassessment of relative values was in order, then fisheries supporters also wondered at the appropriateness of labeling power as progress in view of new technological advances. Was there a need to move quickly in dam development, they asked, when nuclear energy might soon be available? One letter to the *Vancouver Sun* looked to the future in 1958 and judged that "hydro power begins to look mighty old-fashioned." "Or is there a move on foot," continued the anonymous writer, "puzzled," "merely to protect big financial investments already made?" In an inversion of the current rhetoric, the writer concluded that water development "could hold back Canadian progress."[82] One UFAWU local accused General McNaughton of "talking horse and buggy policy in an

79. BCARS, GR 1378, BC Commercial Fisheries Branch, Box 8, File 4, "Preliminary Report on the Fishing Industry of British Columbia (Gordon Commission Study), 1955."
80. BCER CF, "Fish Proving Their Own Case in Fraser River Power Fight," *Columbian*, July 27, 1961.
81. BCER CF, "Fish vs. Power," *Province*, January 21, 1958.
82. BCER CF, letter to editor, signed "puzzled," *Vancouver Sun*, December 8, 1958.

age of Sputniks."[83] Other prominent fisheries advocates such as Roderick Haig-Brown and James Sinclair similarly looked to atomic energy for a future solution.[84] After Dal Grauer, president of BC Electric, gave a speech in 1958 that suggested the revolutionary consequences of energy development in the coming century, John L. Pitman of Coquitlam commented wryly that "We could raise our standard of living so high that we wouldn't have a piece of BC left – it would be sold down the river."[85] Power promoters insisted, in response to such criticism, that water development would create industry and thus the need for future nuclear power development.[86] With or without atomic energy, argued Charles Nash, president of the BCPC, sooner or later all of BC's streams would be needed.[87] Both sides in the debate claimed to embrace the atomic energy future but disagreed over whether it would provide deliverance from the bonds of hydro or merely industrialize the province that much more.

Regional Questions

Any discussion of BC's future, and especially one that put it in such stark and divided light as did the fish vs. power debate, raised questions of self-definition, identity, and the British Columbians' many experiences of regionalism. Because so many of the proposed power projects of the 1950s promised to develop the Interior and spread industry throughout the province, numerous small-town politicians, editors, and boosters seized on power's opportunity and identified a regional interest with dam development. From this perspective, the opposition of fisheries interests to interior projects on the basis of protecting spawning ground habitat appeared to

83. NAC, RG 23, Box 1225, File 726-11-5, part 13, Harold Wilcox, secretary of the New Westminster local of the UFAWU to Angus MacLean, minister of fisheries, December 27, 1957.
84. BCER CF, "Fisherman Urged to Oppose Dams," *Vancouver Sun*, March 28, 1957; "A-Power to End Fish-Power Feud Sinclair Predicts," *Victoria Times*, February 20, 1957.
85. BCER CF, John L. Pitman to editor, *Vancouver Sun*, March 6, 1958.
86. BCER CF, "Moran Dam Delay Loses Atom Plant," *Vancouver News Herald*, January 29, 1957.
87. BCER CF, "River Guards Imperil Salmon," *Vancouver Sun*, February 25, 1958.

be just one more form of metropolitan dominance meted out to an underappreciated and striving hinterland. When fishing interests took a prominent role in discrediting a proposed dam on the Quesnel River in the late 1940s and early 1950s, for example, newspaper editors located in the Interior were quick to identify how these actions benefited "the Coast" against "the Interior."[88] In other words, the fish vs. power debate had a clear regional delineation from an interior vantage point. Saul Rosenberg, representing the Salmon Canners' Operating Committee, responded by pointing out that the fishing industry not only benefited the Coast but also contributed through taxes to the entire province. He sent his views to local newspapers in Prince George, Williams Lake, Quesnel, and Vanderhoof.[89] In a private response, W. L. Griffith, editor of the *Cariboo Observer*, explained that as a former resident of the Coast he understood the importance of the fishery. "However," he continued, "the central interior of this province is a treasure trove that is far beyond the conception of most people residing in BC. It needs, essentially, power to bring it to full production."[90] Implicit in this comment, and in the episode, was the frustration of interior boosters with the slow pace of extending the electrical benefits of the provincial power commission. Against the promise of the late 1940s, when interior boosters and provincial politicians extolled the possibilities of spreading industry and integrating the hinterland, the present paled. "Swivel-chair tacticians in Victoria," complained a *Kamloops Sentinel* editorial in 1953, are more concerned with financing than "imperative needs. . . . Indecision, however, is a poor start for any venture. The plans are there. Let the commission proceed. Now."[91] The Coast – some amalgam of fishing interests, the government, and metropolitan power – acted as the arbitrary and ill-informed force blocking interior aspirations. Throughout

88. Editorials, *Cariboo Observer*, January 25, 1951, February 1, 1951, February 15, 1951, March 1, 1951; BCARS, GR 1378, BC Commercial Fisheries Branch, Box 5, File 5, "Still a Good Case for Hydro," *Cariboo Observer*, April 6, 1951.
89. UBC, Fisheries Association of BC Papers, Box 45, File 11, S. M. Rosenberg, chairman of the Salmon Canners' Operating Committee to editor of the *Cariboo Observer*, February 20, 1951 (copy).
90. UBC, Fisheries Association of BC Papers, Box 45, File 11, W. L. Griffith, editor of *Cariboo Observer* to R. E. Walker, Salmon Canners' Operating Committee, April 3, 1950.
91. BCER CF, "The Glory of Power," *Kamloops Sentinel*, July 10, 1953.

the fish vs. power debate, the anticipation and ambition of interior elites in pressing for development added the aspect of internal regionalism to the discussion.

The politics of water development played on established regional definitions within BC, but also helped to reinforce notions of BC's separateness from without. British Columbians compared their planned advances in dam development or fisheries conservation with those of outside rivals: Canada's industrialized East and the U.S. Pacific Northwest. When McNaughton's diversion plan gained public attention in late 1954, for example, it was routinely linked in public discussion to the St. Lawrence Seaway: as if it were Western Canada's reply to eastern development. There was both rivalry and resentment implicit in the comparison. The view was widely held that "eastern" interests had delayed or denied western development during the war, as in the case of BC Electric's failed expansion bids at Bridge River. Such views grew out of a broader tradition of western regionalism that posited a pernicious power imbalance based on Central Canada's financial dominance and influence in federal affairs, but also gained specific expression in BC political culture of the 1950s. BC's bid to develop the upper Columbia in cooperation with the American Kaiser Corporation gained wide scorn from the provincial press, whereas the federal initiative embodied in McNaughton's diversion proposal received praise.[92] The perceived contrast between the two choices was well summed up in a questioning *Province* editorial headline in 1954, "Power for an Empire, or for peanuts?"[93] Regional interest thus

92. Swainson discusses the press reception of the Kaiser project in *Conflict Over the Columbia*, p. 59. The following is a sampling of the vast outpouring on these issues: BCER CF, "US Interests Would Finance Mica Dam," *Cranbrook Courier*, September 30, 1954; "The Penalty of Mistakes on the Columbia," *Province*, September 24, 1954; "Canada Plans Her Own Columbia Hydro Empire," *Vancouver Sun*, January 30, 1954; "Hydro Men Propose Great River Merger," *Vancouver Sun*, December 31, 1954; "Columbia Diversion May Provide New Irrigation," *Vernon News*, December 30, 1954; "Columbia Power Development Still Big News," *Revelstoke Review*, December 30, 1954; "Ottawa Power Control?" *Columbian*, December 29, 1954; "Columbia Power Plan Staggering in Scope," *Province*, December 29, 1954; "The Wasted Water," *Victoria Times*, December 23, 1954; "Ottawa Moves to Block BC-Kaiser Power Deal," *Province*, December 22, 1954; "Keeping Our Power at Home," *Vancouver Sun*, December 31, 1954.
93. BCER CF, "Power for an Empire or for Peanuts?" *Province*, December 30, 1954.

could adhere to plans proposed from elsewhere so long as they spoke to the perceived interests of British Columbians.

Of course, the praise for McNaughton's plan was also linked to another aspect of BC's identity: its distinction from the U.S. Pacific Northwest. The defensive nationalist overtones of the Columbia Treaty negotiations muddied, to some extent, the active desire of British Columbians to emulate and best their southern neighbors. Harry Warren's statement of the promise of Moran in 1951, for example, explained its potential vastness through continual comparison with American projects: Moran would develop as much capacity as a quarter of all U.S. projects built between 1930 and 1950; the reservoir lake would be longer than Lake Mead behind Hoover dam; Moran would have possibly three times the storage of the Grand Coulee dam; its height would almost equal that of Hoover dam.[94] Warren's claims of BC's coming rise contrasted the poor comparisons of the past. From the early 1930s, when the Columbia River dams gained continental attention, individuals and promoters repeatedly compared BC's *lack* of development with the gains of the Pacific Northwest states, despite a parallel or even superior provincial resource endowment. British Columbians, the claims went, paid more for electricity than Americans but used less of it, commanded fewer dams, and had, as a result, less industry. The spirited defense of Canadian and BC interests contained within McNaughton's diversion plan, however, appeared to provide an opportunity to wrench BC from a position of inferiority and provide a sense of control over American development as never before. When in the late 1950s Columbia negotiations continued to drag, newspapers reveled in the fact that "we can still divert" and that BC "holds the high cards."[95] "The US state department experts," the *Vancouver Sun* reported smugly in 1957, "view this threat with great trepidation."[96] The power of the diversion idea as a combination of national and regional self-assertion gained meaning through the ability to deny the United States its wants.

94. Warren, "National and International," p. 258.
95. BCER CF, "Who is Stalling Downstream Benefits?" *Province*, March 12, 1958; "Sharing Columbia Hydro," *Columbian*, April 28, 1958.
96. BCER CF, "Big Stick Hidden at Columbia Dam Talks," *Vancouver Sun*, May 21, 1957.

When questions turned to the fate of fish under water development, the counterpoint of the U.S. experience also suggested itself. As D. A. McGregor wrote in the *Province* in 1953 concerning the fish – power problem, "The shadow of the Columbia's unhappy fate hangs over the Fraser and adds to the gravity of the Fraser problem."[97] This fact provided cause for self-congratulation at Canadian foresight and raised the question of American indebtedness to Fraser fisheries preservation. A *Province* editorial in 1958 drew a parallel between the Columbia controversy over downstream benefits: "To obtain power the Americans have ruined their Columbia salmon fisheries. To save salmon we have so far resisted power dams on the Fraser. Perhaps we should begin to ask for upstream salmon benefits on the Fraser as a bargaining point for a greater share of downstream benefits on the Columbia."[98] Dam supporters, on the other hand, saw the preservation of Fraser fisheries and the international division of the resource as a scandalous resource giveaway to the Americans. William Ryan, writing in 1959, suggested a cause-and-effect relationship in the building of the Columbia dams and the negotiation of the Pacific Salmon Convention: The first had led to the second.[99] In a 1957 letter to the editor, John Green of Aggasiz asserted that "[W]e are keeping ourselves poor to raise salmon for power rich Pacific Northwest states."[100] General McNaughton, in confidential documents explaining the diversion plan in 1954, expressed a similar frustration at the power of the United States in Fraser fisheries matters: "[V]ery unfortunately, the United States has been permitted to spend money on the remedial measures for the Fraser River slide and the like and we face claims that a servitude has become established."[101] The Columbia fisheries could thus play the role of the cautionary tale and evoke a sense of injustice at the supposed U.S. control of Canadian resources. If regionalism suggested internal coherence against outside forces, it also produced

97. BCER CF, "Fraser and Columbia Power Versus Fish," *Province*, December 11, 1953.
98. BCER CF, "Upstream, Downstream Benefits," *Province*, May 20, 1958.
99. BCER CF, William E. Ryan, "Our Neighbours Always Drive a Hard Bargain," *Province*, March 14, 1959.
100. BCER CF, John Green, "Develop Power Before It's Lost to Fraser Valley," *Province*, December 21, 1957.
101. SFU Archives, W. A. C. Bennett Papers, Box 5, File 4, A. G. L. McNaughton, chairman, Canadian Section, IJC to secretary of treasury board, November 2, 1954 (copy).

division. Although regionalism accented the fish vs. power debate in sometimes surprising ways, it could not overcome this resource conflict or convince British Columbians that all citizens shared a single interest in either fish or power.

Compromises?

The search for solutions to the fish–power problem led to a period of creative problem solving in which both sides of this debate attempted to address the interests of the other. Fisheries supporters sought to promote alternative sites for development on non-salmon-bearing streams. Fisheries scientists studied the passage of fish over high dams and developed fishways to pass low dams. Power promoters looked to the prospects of transferring salmon runs to new areas, to artificial propagation and fish farming. BC Electric funded research on the fish–power problem and examined thermal energy as an alternative to water development in the short term. The Fraser Basin Board integrated different approaches to river management and sought to create a multiple-use plan for Fraser development. All of these solutions seemed to speak to the oft-stated goal of working "in coordination and not in conflict."[102] Yet, to a considerable extent, these solutions did not escape the goals of their makers to grant priority to one resource over the other. The bearers of solutions more frequently sought to position themselves politically than to compromise.

Despite Charles Nash's claim that one day all of BC's rivers would be needed for development, fisheries supporters placed emphasis on searching out and promoting alternatives to the Fraser in the Interior and the North. They believed that if power development occurred elsewhere, the Fraser would be protected. In the 1950s, the IPSFC began a special research program to survey BC waterpowers on non-salmon-bearing streams to provide alternatives to Fraser sites when needed.[103] This defensive strategy accompanied an aggressive political agenda pursued by the organized sections

102. BCER CF, "In Co-ordination, Not in Conflict," *Province*, May 7, 1948.
103. NAC, Pacific Region, RG 23, Vol. 2301, Folder 2, Proceedings of IPSFC meeting, June 20, 1950.

of the fishing industry. These groups began to lobby in favor of a Columbia development program to create the political and economic conditions thought necessary to defend the Fraser. At meetings of the Fisheries Protection Committee in 1958, representatives of different fisheries groups and agencies reflected on the best solution to the Columbia conflict from the perspective of fisheries conservation. The mutual conclusion was to see the McNaughton plan bypassed in favor of a cooperative development with the United States in which some portion of the power created would revert to Canada and be sold by the Provincial Power Commission.[104] A new block of power would thus arrive, minus any diversions and Fraser dams. To support the goal of Columbia development, fisheries groups spoke in favor of international development, encouraged American fisheries groups to lobby on its behalf, and met with American section members of the IJC.[105] The UFAWU went so far as to sign up the support of thirty provincial labor groups, including the BC Federation of Labour and the International Woodworkers of America (IWA) Convention in support of the resolution: "No dams on the Fraser and Development on the Columbia by the BC Power Commission."[106] From this Columbia-centered perspective, even the promised Peace River development was feared initially by fisheries defenders as a distraction from the Columbia development and a possible source of delay that would produce nothing but promotional hype.[107] When it became apparent that the Peace River development might prove feasible after all, fisheries defenders added the Peace to their list of possible "solutions" to damming the Fraser and urged the premier to proceed.[108] Thus behind the complex politics of the Columbia River Treaty was a strong and organized fisheries lobby, pressing for a particular

104. UBC, Fisheries Association of BC Papers, Box 23, File 23-8, Minutes for Fisheries Development and Protection Committee, June 4, 1958; United Fisherman and Allied Worker's Union Papers, Box 138, File 138-3, "Facts on Fish," July 3, 1956.
105. Ibid., Minutes for January 2, 1958.
106. UBC, United Fisherman and Allied Worker's Union Papers, Box 138, File 138-3, Tom Parkin to Walley O'Keefe, April 8, 1958 (copy). In this letter, Parkin explains the success of the resolution.
107. Ibid., Minutes for April 7, 1958.
108. BCARS, GR 1414, Premiers' Papers, Box 37, File 1, J. M. Buchanan, Fisheries Association of BC to premier, October 10, 1957.

development pattern, primarily with a view to sparing the Fraser sockeye. Not surprisingly, General McNaughton appealed on more than one occasion for the fishing industry to refrain from intervening in the Columbia negotiations, lest Canada's position be undermined.[109] Fisheries supporters ignored these requests and continued to state the view that both fish and power could be possible in BC. "Let's have our kilowatts," went a UFAWU pamphlet, "and eat our salmon too."[110]

Power promoters would have replied to this slogan: "Let's have our kilowatts and make salmon adapt to change." While fisheries supporters suggested alternatives in order to keep any development off the river, power promoters adopted the rhetoric of "coordination," "multiple use," and "associated development." Power promoters believed that dams would deliver such enormous wealth to the province that they could not be resisted. They could, however, be planned for and accommodated by those preexisting interests who would be affected. Thus the challenge was to create organizations able to plan multiple use for development, fund research on the passage of fish around dams, and lobby the public about the possible flexibility of salmon, the views of fisheries scientists notwithstanding. That this push for accommodation was substantially one-sided in its intent is apparent in the stated goals of the Fraser River Multiple Use Committee, written to Prime Minister Diefenbaker in 1958, by Committee Chairman and securities dealer J. E. Kania: "It is our belief that [the fish vs. power] conflict is more apparent than real and that it does not constitute a problem without solution. We are of the opinion that the multiple purpose development of the Fraser River would not harm the fisheries but would, on the contrary, assist in the realization of their full potential."[111] When fisheries representation was invited to this committee dominated by engineers, financiers, and professors, committee

---

109. UBC, Fisheries Association of BC Papers, Box 23, File 23-8, Minutes for Fisheries Development and Protection Committee, September 3, 1958; Box 45, File 45-39, "Memorandum of Meeting of General McNaughton with Honourable James Sinclair in Vancouver BC, August 5, 1958."
110. UBC, United Fisherman and Allied Worker's Union Papers, Box 137, File 137-2, pamphlet, "Well I'll Be Dammed: A Fish Story," nd.
111. SFU Archives, W. A. C. Bennett Papers, Box 11, File 3, Kania to Bennett, October 2, 1958 with enclosure, Kania to Prime Minister Diefenbaker, October 2, 1958.

members agreed that the move was "fraught with dangerous possibilities."[112] The committee spoke favorably about the Moran plan, considered calling themselves the Moran Dam Fact-Finding Committee, and included Harry Warren in their membership.[113] "Multiple use" for these advocates was shorthand for "dam development." At another level, BC Electric made efforts to overcome the fish–power problem and its attendant political entanglements by spending its way out of the trouble. In 1956, the corporation organized an interdisciplinary team of researchers at UBC from physics to fisheries biology to "solve" the technical problem. The grant-in-aid of $50,000 was the largest yet received in the area of fisheries at the university and gained considerable publicity.[114] Through an intensified program of science, it was hoped, salmon and dams could be adapted to one another. Power promoters who held a firm belief in the practical capacities of applied science never doubted this outcome. If fish could not be passed around dams, then new spawning grounds could be created artificially, or salmon could be farmed. Happily, reported Russell Potter, a principal of the Moran Development Corporation, in the *BC Professional Engineer* in 1957, a sockeye salmon "has a wonderful homing instinct and will fight to the death to return to his home stream. If it is possible to design a dam so that natural instincts of the fish are exploited in every way, and they are kept away from harm, it should be possible to take them past a dam with little, or no loss, on either of their migrations. This is the principle followed in the design of the Moran dam."[115] If dams could not meet salmon needs, then salmon could meet their own. "I am sure salmon can be re-educated," said Social Credit Mines Minister Kenneth Kiernan in 1960. "We'll be raising salmon the way we raise chickens."[116] "After all," seconded

---

112. UBC, Fisheries Association of BC, Box 45, File 45-34, "Minutes of Fraser River Fact-Finding Committee," April 1, 1958.
113. UBC, Fisheries Association of BC, Box 45, File 45-34, "Minutes of Fraser River Fact-Finding Committee," February 20, 1958; "Minutes of Third Meeting of the Fraser River Multiple Use Committee," March 7, 1958; "Minutes of the Fourth Meeting of the Fraser River Multiple Use Committee," March 15, 1958.
114. This project will be discussed in the next chapter: "The Fraser River Hydro and Fisheries Research Project Final Report." (Vancouver: UBC, 1961).
115. R. E. Potter, "Moran Dam – Fish and Power," p. 24.
116. BCER CF, "We'll Raise Fish as We Do Chickens," *Columbian*, April 9, 1958; "'Educate the Fish' in Fraser – Kiernan," *Province*, April 11, 1960; editorial, *Chilliwack Progress*, March 25, 1960.

Gordon Shrum, chairman of the BC Energy Board, "we don't depend on wild chickens for our eggs or wild buffalo for our meat."[117] Power promoters thus displayed a strong faith in the promise of technology and the pliability of nature; they also had a firm sense of their interest.

The efforts of fisheries supporters and power promoters generally treated coordination as a set of trade-offs, bargains to be struck, sometimes in cooperation, sometimes by establishing priority for one use over the other. The only serious attempt to develop coordination, as both a means to river management and an end, was the Fraser Basin Board, formed after the flood of 1948. This institution joined together civil servants from both the federal and provincial governments and from a wide range of backgrounds, including fisheries, public works, lands, and other relevant ministries and departments. The goal was to overcome problems of jurisdiction and contrasting resource interests in order to evolve a higher form of multiple-use development led by experts from different resource fields, in keeping with such precedents of integrated planning as the TVA. Throughout the late 1940s and 1950s the board worked toward a steady accumulation of studies on possible Fraser River flood-control measures, particularly with a view to dam development. Like a dark horse, the board quietly undertook its tasks, out of the spotlight of media scrutiny and without the self-promotion of the Warren or McNaughton variety. Although fish and power camps railed against one another in public over the Moran and diversion proposals, the board integrated the advice of fisheries representatives with surveys of possible dam sites and their design.

The impetus to overcome barriers of political and sectoral interest, however, could not erase the problems of jurisdiction and the resource priorities of the board's members. From the beginning, the board was tethered by its inability to act: It was an ad hoc body, with only advisory capacity. Unlike formidable river institutions in the United States, such as the Army Corps of Engineers or the Bureau of Reclamation, the board could only study and recommend. The distance between federal and provincial officials and the fact that their primary responsibilities lay elsewhere made the operation

---

117. BCER CF, "Tame Fish, Dam the River – Shrum," *Vancouver Sun*, March 19, 1960.

of the board difficult. The different perspectives brought to bear on the research and planning priorities also created divisions within the institution that limited the ability to act jointly. In 1958, Arthur Paget, BC's water comptroller, wrote to Ray Williston, the minister of Lands and Forests, concerning his experience on the board and his ideas for its redesign. In the future, he argued, the problems to be faced would be largely of an engineering variety and thus the contributions of the federal Department of Fisheries, "whose approach in matters of this kind is essentially negative by reason of their particular interest," could be abandoned. Better, he argued, for the board to hire its own biologists to carry out fisheries studies according to engineering needs than to allow for the interference of the federal department.[118] To some extent, Paget's claim was fair: The Department of Fisheries did wish to block dam projects on the Fraser's main stem, and organizations such as the IPSFC lobbied the federal department to monitor the board's activities.[119] As Tom Reid said on one occasion, although the board was ad hoc, its members "can make themselves very obnoxious, all the same."[120] It should also be remembered, however, that the board members representing other areas of government concern were similarly interested parties. The board's first executive assistant, the engineer Russell Potter, for example, left the board in the mid-1950s to pursue his primary passion: promoting the Moran dam with Harry Warren. Following its 1958 preliminary report, the board increasingly dispensed with fisheries participation in its operations, favoring engineering studies.[121]

Taking the Heat Off the Fraser

The sense of inevitability that surrounded Fraser River dam development in the early 1950s began to crumble in the latter half of the decade. The fish vs. power debate signaled this change by

118. BC Water Management Branch, Department of Lands 'O' File Correspondence, File 0207956, A. F. Paget, comptroller of water rights to Ray Williston, November 25, 1958.
119. NAC, RG 23, Box 1225, File 726-11-5, part 17, Tom Reid to G. R. Clark, deputy minister of fisheries, February 9, 1960.
120. NAC, Pacific Region, RG 23, Vol. 2301, Folder 2, Proceedings of IPSFC meeting, June 20, 1950.
121. Sewell, *Water Management*, especially Chapters 4 and 5.

entertaining discussions from both sides about the possibility for alternatives. By 1957 some of these alternatives gained political momentum. The federal government suggested its commitment to coordinated development on the Columbia by moving negotiations beyond the IJC to the formal diplomatic level. The provincial government promoted the potential of the Peace River as BC's major power source for the future. And BC Electric, judging that the fish–power problem would not be solved quickly, began an expansion of its thermal capacity to bridge the company's power supplies until a major hydro program could be entertained.[122] As the editors of the *Province* put it in 1959, "The projected developments of the Columbia and the Peace have taken the heat off of the Fraser argument."[123]

With neither the federal or provincial government nor BC Electric backing a Fraser River development program, power promoters became increasingly isolated. Although the appearance of the Multiple Use Committee in 1958 seemed to demonstrate the emergence of a new power coalition backing the Moran proposal, the committee failed to raise political or financial support and fell apart after two years. The committee's goal to inspire a royal commission on Fraser River development was rejected out of hand by the federal Conservative government in 1958.[124] General McNaughton's diversion idea similarly lost support. Although the federal government remained publicly committed to the diversion as late as 1958, behind the scenes the attitude was different: As early as 1956, Jean Lesage had advised Premier Bennett that he now judged the diversion unfeasible; henceforth it would be just "a card-in-the-hole for use in international negotiations."[125] Increasingly, the federal government diminished McNaughton's negotiating authority on the

122. BCARS, GR 1427, BC Water Rights Branch, Box 6, File 353, Department of Northern Affairs and National Resources, Water Resources Branch, W. R. D. Sewell, "Prospects of Large Scale Thermal Power Development in the Lower Mainland of BC," April, 1958. Concerning the Moran Development Corporation's attempt to block the company's expansion application before the BC Public Utilities Commission, see BCER CF, "'Hysteria' Charged in Dam Fight," *Vancouver Sun*, November 15, 1957.
123. BCER CF, "Fraser: Fish, Floods and the Future," February 13, 1959; on the postponement of a Fraser project in light of the shift of policy, see "BC Electric gives up Fraser Power Plan, Ends Bitter Fight," *Financial Post*, May 18, 1957.
124. NAC, RG 23, Box 1225, File 726-11-5, part 14, Kania to MacLean, November 14, 1958; MacLean to Kania, December 3, 1958.
125. SFU Archives, W. A. C. Bennett Papers, Box 11, File 3, "Federal–Provincial Discussion on International Waters," July 4, 1956.

Columbia file, while Premier Bennett tried to force McNaughton to admit that the diversion would not proceed. BC Electric's shift to thermal power also spoke to the company's view that the diversion plan had run into too many political obstacles.

Although the conditions that would have allowed for major developments on the Fraser began to change in the late 1950s, the fish vs. power debate did not subside. In part, this was because there was no end to the Moran and diversion promotions: The Multiple Use Committee pledged to end the fisheries hold on the river, the federal government did not back away publicly from the diversion plan, and General McNaughton became increasingly strident in his advocacy of a diversion as its likelihood diminished, declaring before a committee of the House of Commons in 1957 that "We are entirely masters of our own destiny."[126] Added to these continuing questions was the looming possibility that the Fraser Basin Board might convince the federal and provincial governments to proceed with flood-control dams. The board's 1958 preliminary engineering report laid out a blueprint for development, the so-called System A plan: a series of dams in upper-basin tributaries that would help to ward off the threat of floods and pay for this service by creating small blocks of power in some of the multiple-use structures. There would be power and storage dams north of Prince George on the McGregor River and at Olson Creek; farther south, two dams would be placed on the Cariboo River, a tributary of the Quesnel, and another five on the Clearwater, feeding the North Thompson. Several dams were envisioned with the option of another on Stuart Lake if the fisheries problem were not too severe (see Map 9).[127] The report received favorable comment from the provincial government in a major speech by Ray Williston on hydro policy in 1959.[128] Although the Department of Fisheries expressed pleasure at the apparent efforts of the board to avoid damming spawning grounds in their plans, other fisheries

126. BCARS, GR 1427, BC Water Management Branch, Box 3, File 92, "House of Commons First Session – Twenty Third Parliament 1957, Standing Committee on External Affairs, Minutes and Proceedings, Statement by General A. G. McNaughton, Chairman, Canadians Section, International Joint Commission" (Ottawa, 1958), p. 294.

127. "Preliminary Report of the Fraser River Board," Victoria, 1958.

128. NAC, RG 23, Vol. 1225, File 726-11-5[15], Ray Williston, "Hydro-Electric Power in Canada," an address by the Hon. R. G. Williston, minister of lands and forests, delivered during the Debate on the Speech from the Throne, Legislative Assembly, January 27, 1959.

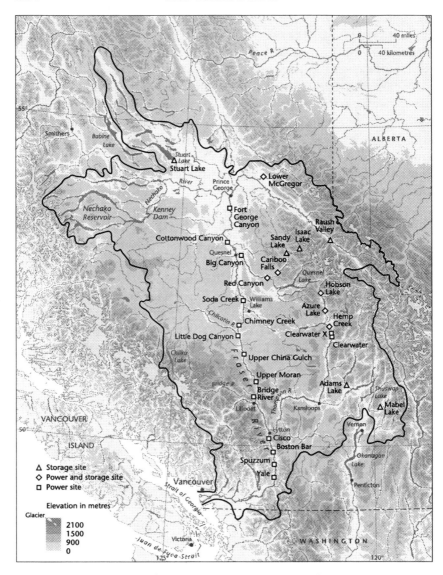

Map 9. Fraser Basin Board power site investigations, 1958.
(*Source:* Fraser River Board, *Preliminary Report on Flood Control.*)

organizations made their fears known: The UFAWU, the Prince Rupert Fisherman's Cooperative, and the Native Brotherhood all lobbied the federal government to block the proposed dams.[129] "Nowhere can we find reference to the problem of food supply for the Interior Indians," wrote Ed Nahanee of the Native Brotherhood, in reference to the Fraser Board's report, to Minister Angus MacLean in 1959.[130] The System A proposal would eventually unravel in the mid-1960s because of fisheries concerns and the high costs of flood-control dams relative to diking in the Lower Fraser Valley. But before that time, the board's activities kept the controversy of fish and dams on the Fraser alive.

Nor did fisheries defenders feel safe to rest on their laurels in the late 1950s, despite a number of strategic victories: The fish vs. power debate thrived on a paranoid style of politics. Indeed, one federal economist described the fish vs. power debate as its own "cold war."[131] Ever since the late 1940s, when fisheries supporters found their interests trampled on in the Alcan case, the potential of a repeat remained a concern. After BC Electric committed itself to fund studies of the fish–power problem at UBC in 1956, for example, IPSFC Commissioner and timber and cannery executive H. R. Macmillan wrote to UBC fisheries biologist Peter Larkin: "Peter: Don't let them put the fish people to sleep. Their tactic could be to act like they don't need [the] Fraser, then find [the] Columbia too slow and costly. Put on a few brown outs, say the only cure would be couple of quick low dams on [the] Fraser. And the girl would be only a weeny teeny bit pregnant."[132] In response to the Fraser Basin Board's conciliatory approach to the fish–power problem that sought sites beyond harm to fish, Tom Reid suspected nefarious intentions, or at least a historical logic: Once these dams appeared

---

129. NAC, RG 23, Vol. 1225, File 726-11-5, part 15, "Memorandum: Fisheries Aspects of the Preliminary Report on Flood Control and Hydro-Electric Power in the Fraser River Basin," A. L. Pritchard, March 19, 1959; UFAWU Press Release, December 19, 1958; part 19, K. F. Harding, secretary for the board of directors, Prince Rupert Fisherman's Cooperative Association to MacLean, June 8, 1962.

130. NAC, RG 23, Vol. 1225, File 726-11-5, part 15, Ed Nahanee, secretary–business agent, Native Brotherhood of BC to Angus MacLean, minister of fisheries, January 12, 1959; MacLean to Nahanee, January 20, 1959 (copy).

131. NAC, RG 23, Vol. 1230, File 726-11-13, I.S. MacArthur, economic staff, to deputy minister of fisheries, December 21, 1956.

132. Quoted in Drushka, *HR: A Biography*, p. 305.

on the Fraser, others would inevitably follow.[133] Tom Parkin, a
publicist for the UFAWU, similarly charged that the Fraser Basin
Board's plans were a "backdoor" for development to enter the
Fraser Basin.[134] In 1958, well after the diversion had commanded
central attention in Columbia discussions, the Fisheries Protection
and Development Committee was planning means to discredit it.
Even the Peace development, thought to be a partial solution to
the Fraser fish–power problem, could be viewed, from a certain
perspective, as a looming threat. What if the real intention were
to divert the Peace into the Fraser? Or what if upper Fraser tributaries were diverted into the Parsnip River to provide extra flow for
a Peace dam? Not only would the Fraser's hydrology be affected,
but also, as one UBC biologist argued, there was a risk of releasing
new fish predators into the Fraser Basin from this northern watershed.[135] Although some fisheries supporters became increasingly
triumphant in public when signs of the Peace and Columbia programs appeared to clear the way for fisheries conservation on the
Fraser, others refused to accept that the threat had passed or could
be ruled out in the future.

That this suspicion had foundation became apparent in the early
1970s. As the development program on the Peace and Columbia
came to completion in the late 1960s, the BC Energy Board, established under the chairmanship of Gordon Shrum in 1959 to
advise the province on energy policy, turned again to the question
of power development on the Fraser. Shrum's views on the fish–
power problem were put bluntly to the retired canner Henry Doyle
in 1956: "I think even the most ardent commercial fisherman in
British Columbia realizes that eventually the Fraser River will have
to be used for power."[136] Over a decade later those views appeared
not to have changed. Under Shrum's direction, the BC Energy Board
launched a major feasibility study to examine the possibility of

---

133. NAC, RG 23, Box 1225, File 726-11-5, part 17, Reid to G. R. Clark, deputy minister
of fisheries, February 9, 1960.
134. BCER CF, "Power Without Harm to Fish 'Possible,'" *Province*, March 14, 1959.
135. BCER CF, "Professor Says Peace Dam Threat to Fish," *Vancouver Sun*, November
14, 1958; UBC, Fisheries Association of BC Papers, Box 45, File 45-16, C. C. Lindsey,
"Possible Effects of Water Diversions on Fish Distribution in British Columbia," nd.
136. UBC, Doyle, Henry Papers, Box 1, File 1-8, Shrum to Doyle, May 9, 1956.

developing the Moran dam.[137] Helicopters hung ominously above
the site in 1971, taking drilling samples to test the foundations for a
major dam footing, while a team of biologists, economists, and engineers considered costs and benefits.[138] "Gathered in offices in the
BC Hydro building and at the universities," wrote one reporter in
the idiom of a new age, "the high priests of our technocratic society,
engineers and scientists, are performing mystic rites to determine
whether or not to dam the Fraser River." The priests, the reporter
continued, were trying to turn the river from a "strong brown god"
into an "electric generator."[139] The public response to the plan was
overwhelmingly negative.[140] In addition to the fisheries protest, an
emergent environmental movement expressed astonishment that
the pristine river would be thus violated. In 1970, E. H. Vernon, a
member of the BC Fish and Wildlife Branch, evoked the sensibili-
ties of this reaction in a private letter considering the development
of the Fraser: "Any element in our society that grows by doubling
every 10 years quickly must reach the stage of being ridiculous or
impossible or both.... Must we turn the world into a Los Angeles
before we stop talking stupidly of unending growth?"[141] Premier
Bennett, quick to sense the direction of discussion, disavowed the
plan. Social Credit highways minister, the flamboyant evangelist
"Flyin'" Phil Gaglardi, was one of its few defenders. When one
journalist prodded him, saying that Bennett had claimed that no
dam would be built on the Fraser while he was premier, Gaglardi
replied, "Yes, it won't be built while he's Premier – but you've got

137. Minutes of meetings, memos, and correspondence concerning this prospect are contained in BCARS, GR 442, BC Energy Board. UBC, Warren Family Papers, Box 4, File
12, "Moran Dam Study Asked for Fraser River," *Vancouver Sun*, January 28, 1971.
138. Bocking, *Mighty River*, Chapter 6.
139. UBC, Fisheries Association of BC Papers, Box 46, File 11, "High Priests vs. the Raging
Fraser," *Vancouver Sun*, June 1, 1971.
140. UBC, Fisheries Association of BC Papers, Box 46, File 11, "More and More on
Moran," *Province*, March 29, 1972; "Ottawa Holds Dam Veto: Davis," *Province*,
April 26, 1972; "Now Why Would He Go to All of that 'Dam' Trouble," *Columbian*,
February 12, 1972; "Haig-Brown Raps Moran Dam Idea," *Sun*, April 10, 1972; "A
Peaceful Corner of the World," *Province*, May 29, 1971; "The Great Demand . . .
Electricity," *Province*, February 17, 1971.
141. UBC, Fisheries Association of BC Papers, E. H. Vernon, chief Fisheries Management,
BC Fish and Wildlife Branch, to K. M. Campbell, secretary manager of the Fisheries
Association of BC, November 17, 1970.

to remember he's 71 years old now."[142] Within months, Bennett's government would go down to defeat to the New Democratic Party under Dave Barrett, and with that, BC's big dam era would come to a close. In his memoirs, Gordon Shrum threw up his hands and said that he had "become resigned that the Fraser River fish/power problem will not be resolved on a dollars-and-cents basis. It is mainly a political issue."[143]

Conclusion

The fish–power problem on the Fraser River in the 1950s forced a new politics of electrical development in BC. Although numerous interests and governments sought to create political programs to develop the river and coalitions to sway public support, fisheries interests managed to hold off dam projects until new alternatives became possible. Those alternatives on the Peace and Columbia came about in part because of the political pressure to conserve the Fraser as salmon spawning habitat. The Peace and Columbia development agenda thus operated as both a political outcome of, and a solution to, the fish vs. power conflict on the Fraser.

The fish vs. power debate was transnational. At one level, the American case provided an analogy of possible futures. The declines in Columbia salmon stocks after main-stem dam development in the 1930s shadowed debate and discussion on the Fraser. But the transnational connection was also direct and political. The United States held an interest in the Fraser River fishery under the Pacific Salmon Convention. If dam development went ahead on the Fraser, American fishing interests would be affected. The United States also held a strong interest in developing complementary projects on the upper Columbia in order to benefit American power production downstream. Fraser River projects threatened the feasibility or attractiveness of competing Columbia River proposals; and McNaughton's diversion scheme augured worse. Thus

142. UBC Special Collections and Archives, Fisheries Association of BC Papers, Box 46, File 11, "More an' More Opinions Flow from BC Ministers," *Vancouver Sun*, March 29, 1972.
143. Shrum with Stursberg, *Gordon Shrum*, p. 131.

Canadian–U.S. disputes over the Columbia River in the 1950s always implied the Fraser. One river suggested the place and promise of the other.

Ambitious provincial development plans could avoid the Fraser River in the 1950s because there were alternatives on the horizon. Provincial development plans for the Peace River first appeared as a negotiating ploy in the debate with the United States over downstream benefits on the Columbia. But after the signing of the Columbia River Treaty and negotiating the end of the federal ban on power exports, the provincial government seized the opportunity to develop two large rivers at once. As a consequence, the Columbia would proceed, as well as a major dam project on the northern Peace River under the authority of BC Hydro, the new provincially owned corporation. The province thereby secured its power supply, spurred economic development, and tied future hydro policy to a two-pronged strategy: meeting provincial needs and leveraging its own position in emerging continental energy markets.

The content of the fish vs. power debate revealed its context. It spoke to British Columbians' regional aspirations and prejudices, their dreams for the future as well as their fears. It was not an environmental debate in a modern sense of that term: Its major disputants spoke a utilitarian idiom, not a romantic or ecological language; resources were the objects of concern, not wilderness or ecosystems. However, it would be exaggerating the point, as some have done, to assume that few tried to question or criticize unfettered resource and river development in BC in the 1950s.[144] A dispute dominated by utilitarian values did not preclude the consideration of sentiment, tradition, and the needs of others. In the view of some of its participants, the fish vs. power debate was unpleasantly robust.

144. Wilson, *Talk and Log*, pp. 100–1.

7

The Politics of Science

Before a gathering of business people in 1960, UBC fisheries biologist Peter Larkin reflected on the dynamics of the fish vs. power debate and broke the dispute into four "technical ingredients": "the fish, fisheries biologists, engineers and dams." With self-deprecation, Larkin described the purpose of the fisheries biologist thus[1]:

The first characteristic of the fisheries biologist must be slipperiness. Recognizing that fish and their environment are variable, and that even with the best of observations, he has only a general understanding of what's going on, he is forced to approach every problem with a becoming caution. Things are never so; they seem to be, are apparently, they are indicative, it is suggested, maybe they are true. And always from a biologist expect lots of adjectives and adverbs, slightly, moderately, reasonably, average. Fishery biology is largely the art of saying 'probably' in 1000 ways.

If a fisheries biologist is known as an expert, it is *probably* because he says nothing or because everything he says can be construed as a completely satisfactory prediction regardless of what happens.

The *poor* fisheries biologist on the other hand is characterized by his over-confidence. Fancying himself as something of a jet age scientist, and feeling compelled by our scientifically minded society to put up or shut up – he recklessly tries to put up – promising the moon, hoping memories are short, and looking for scapegoats at the hour of disenchantment. . . .

Confronted with the fish–power problem the fisheries biologist can call on a fairly healthy experience, can promise to do his best, largely playing by ear, making ad hoc arrangements and as the English say 'muddling

1. BCARS, GR 442, BC Energy Board Papers, Box 52, Peter Larkin, "Fisheries and Water Resources Development," Pacific Northwest Trade Association, Sun Valley, Idaho, September 26, 1960.

through.' To his critics he can always throw out the challenge 'let's see you do any better.'

This tongue-in-cheek assessment came from one of the leading fisheries biologists involved in the fish–power discussions of the 1950s. A Rhodes Scholar, an ecologist trained under Charles Elton, a professor in the emerging Institute of Fisheries at UBC, a veteran of the BC Fish and Game Branch, and a future member of the FRBC, Larkin knew whereof he spoke. He had conducted impact assessments of the Alcan project in the late 1940s, written programmatic statements on the best means for the university to respond to the fish–power problem, and led a number of studies in search of a solution.[2] Far from being slippery, his views were remarkably frank and, as we shall see, perceptive.

Fisheries scientists played a prominent role in the fish vs. power debate. They studied the effects of dams on fish, launched research programs in the areas of basic and applied biology, advised politicians, joined lobby groups, and spoke to the press. In a variety of ways, they changed the public profile of science as a disinterested pursuit, faced questions of the boundaries between common and scientific knowledge and debated their appropriate mission. In turn, the politics of fish vs. power shaped science. It directed institutional research agendas, had an impact on individual careers, and produced a new vision of salmon and their environmental limits. This chapter seeks to explain the many ways in which scientists engaged in the public debate over the environment and how that intervention changed their knowledge as well as their pursuit of knowledge.

In Chapter 3, on the remaking of Hells Gate, the analysis sought to evaluate the intersections of research practice and institutional and national politics. The role of fisheries scientists in the making and resolution of disputes pointed to the importance of science more generally in the debate over fish and power. This chapter follows that earlier lead, focusing less on questions of practice and the substantive content of science and more on the problems of institutional research patterns and the politics of scientific authority.

2. Crowcroft, *Elton's Ecologists*, p. 49; UBC, Finding Aid, Peter Larkin Papers, biographical statement.

Drawing on sociological literature concerning the politics of "big science" and "invisible colleges," this chapter considers the changing institutional patterns of fisheries research in BC and the initially elevated and later threatened status of fisheries biologists in public discourse.[3] Two key themes emerge: First, that institutions underwent a period of remarkable growth, marked by increased funding, industrial, and political influence, and a shift to greater cooperation within and between institutions; and second, that the status and authority of fisheries science was made and remade by supporters and detractors of this field. These two themes were linked and inseparable. The rising authority of fisheries science in the early 1950s set the conditions for the expansion of institutions and research; the reaction against the results of this research aimed squarely at the reputation of fisheries science; in turn, fisheries scientists defended their work by bolstering their own authority and questioning that of others. Science changed the fish vs. power debate, and in turn the debate changed science.

"Let the Experts Decide"

Fisheries scientists began the 1950s in a position of unprecedented authority. Public commentators marveled at the power of expertise to overcome problems of resource conflict; editorialists and politicians praised scientists' abilities to conquer the unknown at Hells Gate; and new means of institutional and financial support appeared. Fisheries scientists enjoyed heightened prestige, but felt challenged and threatened by the burdens of high expectations. Their new status rested on assumptions that could not last.

Whereas before the 1940s, scientists played a more peripheral role in public debates over resource development in BC, in the postwar period they commanded the role of experts in public discourse. Representations of scientists as experts were but one expression of a more general shift in attitudes toward science and institutionalized authority emerging from the war years. Science and scientists

3. The classic texts on "big science" and "invisible colleges" are De Solla Price, *Little Science, Big Science* and Crane, *Invisible Colleges*. I have also found useful a collection of essays reflecting on the heuristic of big science in concrete empirical investigations: Galison and Hevly, *Big Science*.

were deemed to be impartial adjudicators and subtle blenders of nature and technology; they were granted a new authority to decide the best means of resource development for the general interest. On the issue of fish and power, instructed a *Province* editorial in 1949, "Let the Experts Decide."[4] Editors of the *Columbian* in the same year judged the fish–power issue to be too complex for "laymen" and asked for an expert solution.[5] Advocates of different persuasions – from nature writer Roderick Haig-Brown to the most aggressive dam promoters – shared the belief that many complex political problems in water development could be solved through the application of expertise.[6] The headline of the *Columbian*'s lead editorial in 1960 – "Let's Leave BC Power Fate to Experts – Not Politicians" – expressed the sentiment that expertise could overcome the vagaries of interest and political expediency.[7] Such views envisioned a peculiar democratic responsibility for scientists *cum* experts as the scientific–moral conscience in matters of social and environmental planning.

Local conditions also influenced ideas of the place of science and scientists in the emerging problems of fish and power. The accomplishments of fisheries scientists at Hells Gate convinced some scientists and editorialists that the looming problem of fish and dams could be overcome by the extension of existing knowledge. "In view of the inevitable encounter between dams and fish on the Fraser," wrote Richard Van Cleve, chief biologist of the IPSFC in 1947, "it is most fortunate that much of the work accomplished so far by the staff of the International Pacific Salmon Fisheries Commission has been devoted to the study of the effects of obstructions on migratory fish." The Hells Gate episode, he continued, provided a "pilot experiment" in the future of fish and dams.[8] "Fishery engineers have learned a lot about the practical side of fish conservation at Hell's Gate," stated an editorial, echoing Van Cleve's argument,

4. BCER CF, *Vancouver Sun*, January 27, 1949.
5. BCER CF, "Hydro and Salmon," *Columbian*, January 21, 1949.
6. "BC Seeks Way to Waterpower Without Sacrifice of Salmon," *Pacific Fisherman* 49(5) (April 1951).
7. BCER CF, "Let's Leave BC Power Fate to Experts – Not Politicians," *Columbian*, February 13, 1960.
8. UWA, Van Cleve, Richard Papers, "Paper Presented at Symposium of the Western Division Association of Fish and [?] San Diego, June 18, 1947."

in the *News Herald* in 1949. "The lessons learned are now available to be applied generally."[9] Throughout the 1950s Hells Gate played a symbolic function representing past accomplishment and present readiness in fish–dam cases. In 1953, during the construction of the Seton-Bridge River project, a *Province* journalist suggested that no harm would be done to the Seton fish runs because the project's fishways had "already been proven a success at Hell's Gate, Farewell Canyon and other danger spots in the Fraser system."[10]

Besides this celebrated example, American development on the Columbia provided numerous opportunities for comparison and optimism. Harry Warren and his fellow Moran dam promoter Russell Potter both claimed in numerous public engagements that American advances had solved the problem once and for all.[11] Val Gwyther, a Vancouver engineer and dam enthusiast, claimed that American advances in artificial propagation offered a new opportunity to maximize watershed values.[12] Numerous scientists and government officials visited the Columbia during the 1950s, hoping to learn how BC could profit from the American example.[13] Rather than question dam development, argued Alcan Vice President McNeely Dubose, scientists should look on fish–dam cases as providing new natural "laboratories" for their scientific curiosities.[14] Such pronouncements could sometimes tweak the pride of even the most cautious fisheries scientist. In 1956, in a planning meeting considering the Moran dam proposal, IPSFC Director Loyd Royal said that in earlier times there could be no scientific answer to the problem of fish passing high dams, but in the current fisheries science "renaissance" he felt that the Moran project should be given "every consideration."[15]

9. BCER CF, "Hydro and Fisheries," *Vancouver News Herald*, November 22, 1949.
10. BCER CF, Norm Hacking, "Afloat and Ashore," *Province*, November 25, 1953.
11. Potter, "Moran Dam"; "Moran Dam and the Fraser River," *Engineering Journal* 43(12) (1960): 43–6.
12. Gwyther, "Multiple Purpose Development of the Fraser River Basin."
13. BCARS, Box 4, File 6, George Alexander to master fish warden, Fish Commission, Oregon, February 9, 1951. Alexander's request for information was one example of a stream of correspondence and visits from BC and federal officials.
14. NAC, RG 23, Vol. 1822, File 726-11-6, part 1, Milo Bell to G. R. Clark, October 14, 1949. Bell reports Dubose's statement.
15. BCARS, GR 1118, BC Marine Resources Branch, Box 3, File 1, "Notes on Meeting with Moran Power Development Ltd. Held in the Offices of the Chief Supervisor of Fisheries on May 24, 1956 at 11:00 AM."

Although some scientists embraced public enthusiasm and spoke of their capacity to solve the fish–power problem, many more tried to explain the nature of the challenge. There were no miracles, they argued, only messy problems without clear solutions. At a meeting of the IPSFC with representatives of the fishing industry in 1950, for example, Milo Bell, one of the designers of the Hells Gate fishways with wide experience on Columbia dams, sought to instill in the audience respect for the lack of knowledge of salmon tolerance to obstructions. "[I]f I told you how little we knew when we built the Hell's Gate fishways," he said, "... you might not have given us authorization to build."[16] At a BC Natural Resources Conference in 1953, Bell and others argued to the mixed audience of scientists, government officials, and industrial representatives that no solutions existed to the fish–power problem.[17] Many of the potential problems had yet to be defined. "[T]he biologist is working with all of the complexities of a living creature," said J. R. Brett of the FRBC, "not wholly unpredictable in behaviour, but far from a mechanical horse."[18]

Recent experience suggested the need for caution. Despite the sanguine press reception of the Columbia dam projects and their fishways, fisheries scientists working on Columbia River passage problems frankly admitted the level of their ignorance in the 1950s. The fishways were not solutions so much as they were experiments in action. Harlan Holmes, a Stanford biologist involved with the design and construction of the Bonneville dam fishways, recalled that the fishways had "evolved" as the construction of the dam proceeded, with changes from "day-to-day."[19] Much of the research of the U.S. Army Corps of Engineers on fishways after the 1930s sought to determine basic questions such as how many fish survived their passage over dams. Whether the fishways actually worked with any consistency was still a matter of debate and

16. NAC, Pacific Region, RG 23, Vol. 2301, Folder 2, Record of Proceedings of IPSFC, June 20, 1950.
17. Clay, "Problems Associated with Upstream Migration"; Jackson, "Measurement of Losses of Fingerling Salmon"; Brett, "The Nature of the Biological Problem"; Bell, "Fisheries Research at High Dams."
18. Brett, "The Nature of the Biological Problem," p. 98.
19. Quoted in Mighetto and Ebel, *Saving the Salmon*, p. 55.

research.[20] In Canada, the looming threat of dam projects on the Nechako River in the late 1940s forced Canadian fisheries scientists to admit that they did not yet have even an accurate sense of the scale of the annual migrations to this river, let alone clear ideas about how best to perpetuate runs with the addition of dams.

The problems involved were fundamental. Dam development posed a variety of as yet unanswerable questions: What would happen to salmon, of different stocks and species, when dams from 100 to 700 ft in height were placed in their migratory path? Could salmon climb ladders around these dams? Could they pass only one dam or many dams? Would they find the reservoirs behind dams disorienting? What would happen to fry migrating to the ocean? How would dams affect salmon behavior, physiology, and ecology?

These were questions asked of a science that had but fifty years experience on the BC coast, that had determined in only the past forty years that salmon return to their natal streams to spawn, and that had just recently concluded that obstructions at Hells Gate proved injurious to migrating salmon. In 1953 J. R. Brett instructed thus[21]:

It is useless to search the biological literature for the answer [to the fish–power problem], though much of the background and methods of study will be found there. The fish have not been examined sufficiently under conditions imposed by large reservoirs, submerged outlets, rapid fluctuations in flow, and catastrophic pressure changes, to provide answers. Nor is there any good knowledge of the physical conditions which the fish actually face. No excuses are necessary for this lack of knowledge. The changed water conditions are a product of our times. Solutions to the problems created are the task of our times.

Institutional Expansion

Searching for solutions required more than new ideas. It also required an enhanced institutional framework and greater levels of cooperative research with a view to a coherent provincial research

20. Ibid., pp. 103–7. 21. Brett, "The Nature of the Biological Problem," pp. 96–7.

program. In the postwar period, research agencies expanded and gained greater funding. Formerly distinct agencies overlapped in new ways. A framework amenable to larger cooperative research emerged, partly in response to the politics of fish and power. Previously the FRBC had been the dominant research institution, but it now shared pride of place in advanced fisheries research with the IPSFC. Both agencies cooperated in environmental assessments of water-development projects and sponsored extensive research programs.[22] Beyond the FRBC, the federal Department of Fisheries maintained a source of expertise in matters of habitat restoration in the Fish Culture and Development Branch, an agency devoted in part to assessing dam projects and devising fishways. The previously important provincial Department of Fisheries that had spearheaded salmon biology at the beginning of the century waned in influence, while the provincial Fish and Wildlife Branch gained new importance. Unlike the industry-oriented Fisheries Branch, the Fish and Wildlife Branch focused on questions of the sports fishery and recruited a new generation of fisheries scientists to implement programs and study resource problems.[23] At UBC, a small biology research program expanded and gained a special institute devoted to fisheries. Continuing the previous links between the department and the government research agencies, the UBC Department of Zoology drew on government scientists to teach courses, sent students to field schools at the FRBC station in Nanaimo, hired government staff, and in turn lost scientists and sent former students to the research agencies in a developing pattern of institutional cross fertilization.[24] Peter Larkin, for example, began to teach in the UBC department while an employee of the provincial Fish and Game Branch in the 1950s, became a full-fledged member of the department in the mid-1950s and later joined the FRBC in the mid-1960s, only to return to the university 3 years later. As in other

22. Johnstone, *The Aquatic Explorers*, p. 200.
23. BCARS, GR 1027, BC Fish and Wildlife Branch, Box 1, File 12, J. Hatter, "The Function of the Provincial Wildlife Biologist," paper presented to the first meeting of the Canadian Wildlife Biologists, Ottawa, January 20–1, 1958.
24. UBC, Larkin, Peter Papers, Box 11, File 11-1, W. A. Clemens, "A Review of Work in Fisheries at the University of British Columbia" (stamped "received, January 31, 1952"); W. A. Clemens, *Education and Fish*, pp. 56, 65–6; Clemens, "Some Historical Aspects."

jurisdictions like Ontario and the U.S. Pacific Northwest, fisheries research operated with porous boundaries between state agencies and the academy.[25]

Industrial and client groups exerted a significant influence over the course of research conducted at these institutions. The FRBC, although a major developer of basic research, also pursued a range of applied problems to enhance fisheries productivity and the commercial development of fish products.[26] The IPSFC incorporated industrial advice into its corporate structure through advisory committees and maintained a close liaison with fisheries groups over political problems of habitat protection. The Fish and Wildlife Branch organized some of the most active wilderness protection groups during the 1940s and 1950s and kept in close touch with sports fishers and recreational users.[27] The UBC department emerged with the support of the fish-processing industry. In 1945, BC Packers made a $45,000 gift to help establish an Institute of Fisheries; in the late 1940s and 1950s, all of the major fish-processing firms as well as the UFAWU provided funds for operations, a library, new staff, and student scholarships; and H. R. MacMillan made annual donations and major one-time contributions in the aid of hiring new staff and supporting research.[28] MacMillan's individual contributions arose from his growing interest in fisheries biology spurred in part by his professional experience and appointment to the IPSFC in 1952, but also from his conviction, stated to J. M. Buchanan of the Vancouver Foundation in 1959, that "A strong Fisheries Institute here, staffed by able men of wide background, will assist in keeping the truth before the public.... Work of this nature will enable the Institute to hold a higher grade of men than otherwise, and will also build

25. Bocking, "Fishing the Inland Seas"; idem, *Ecologists and Environmental Politics*, Chapter 7, "Ecology and the Ontario Fisheries"; Mighetto and Ebel, *Saving the Salmon*.
26. Johnstone, *Aquatic Explorers*.   27. Wilson, *Talk and Log*, pp. 98–9.
28. UBC, Larkin, Peter A. Papers, Box 2, File 1, Paul Bullen, UBC accountant to G. Peter Kaye, executive director, Vancouver Foundation, July 8, 1963 (copy); Institute of Fisheries, "H. R. MacMillan Expeditions Grant," 1959; File 2, "Summary of Circumstances Concerning H. R. MacMillan Expeditions Grant," nd; "Report on H. R. MacMillan Grants for Fisheries Work at the University of British Columbia," nd; Box 11, File 11-1, W. A. Clemens, "A Review of the Work in Fisheries at the University of British Columbia" (stamped "received January 31, 1952"); UBC, Fisheries Association of BC Papers, Box 23, File 23-10, "The Institute of Fisheries University of British Columbia," nd; Drushka, *HR: A Biography*, pp. 256, 351–2.

for the Institute a broad international reputation.... When such men speak in the years to come concerning factors that might be detrimental to British Columbia fisheries, their opinions will have earned respect here and elsewhere."[29] MacMillan's particular interest in supporting fisheries science as a means of defending the fisheries was but one example of a broader trend: Industrial support fed the growth of fisheries science in the postwar period with a view to sustainability and commercial growth.[30]

The expansion of research funding for fisheries biology in BC gained impetus from the fish–power problem. MacMillan was not the only interest to support research with a view to political or practical outcomes. In 1956, J. M. Buchanan, chair of the Fisheries Association of BC, pressed federal Fisheries Minister James Sinclair to double the personnel of the Fish Culture and Development Branch in order to speed up investigations of power projects. "We feel it vitally important," he wrote, "that under no circumstances should the industry appear as an obstructionist to progress." Speeding research, he implied, would aid this goal.[31] From the other side of the debate, BC Electric also sponsored a major cooperative study at UBC between engineers and biologists starting in 1956 with the particular goal of considering the feasibility of low-level dams on the Fraser. Its grant of $50,000 was the largest yet received by fisheries biologists at the university.[32] At the federal level, spending on fish–power research was estimated at $60,000 annually after 1949.[33] Over the 1950s, all research agencies in BC devoted just under $5 million to fish–power research. To put that figure in context, the Hells Gate fishways program, including capital costs, reached just over $1.3 million.[34] By way of comparison, fish–power

29. UBC, Larkin, Peter A. Papers, Box 2, File 2, MacMillan to J. M. Buchanan, secretary, Vancouver Foundation, April 24, 1959 (copy).
30. Rajala, *Clearcutting the Pacific Rain Forest*, pp. 70–1. Rajala notes the drive for expanded forestry research at UBC during this same period, propelled by industrial needs.
31. UBC, Fisheries Association of BC Papers, Box 75, File 1, J. M. Buchanan to Sinclair, August 21, 1956 (copy).
32. UBC, Fraser River Hydro and Fisheries Research Project Papers, File 3, E. S. Pretious, L. R. Kersey, and G. P. Contractor, "Fish Protection and Power Development on the Fraser River" (UBC, February, 1957).
33. NAC, RG 23, Vol. 1230, File 726-11-14, part 2, C. H. Clay to Whitmore, May 15, 1958.
34. BCARS, GR 442, Box 52, "Summaries of Research on the Fish-Power Problem and Related Work by Fisheries Agencies in British Columbia," prepared by the research

research on the Columbia in the 1950s carried out by the U.S. Army Corps of Engineers, U.S. Fish and Wildlife Service, and the states of Washington and Oregon more than doubled the Canadian research in terms of dollars spent.[35] Thus, although fish–power research spending in BC dwarfed past efforts, it did not nearly match the American program, which was driven by larger budgets and more immediate problems. Increased funds nevertheless allowed for new research projects and a reorganized institutional effort.

Research Agendas and Cooperation

In response to the public call for intensified fish–power research, fisheries scientists began to develop a focused research agenda. They identified past accomplishments and present lacunae. They imagined future directions. Leaders in research advocated a cooperative program and organized areas for team research. In different research agencies, scientists attempted to set out the particular contribution that they could make to solve the fish–power problem.

In every major programmatic statement on fish–power research needs in the 1950s, fisheries scientists emphasized the importance of fundamental research. Fish–power research, they argued, must not be assumed to be a technical exercise, but a problem of basic science. Federal fisheries scientist J. R. Brett put the point succinctly in a widely distributed paper in 1956[36]:

Since the problem is both multiple and complex, no delusions should be entertained concerning the possibility of some quick or simple solution. Any new mechanical contrivance expected to aid salmon at some point in their migration will create new biological problems. *It is the lack of knowledge of salmon that is the great handicap.* This handicap can only

subcommittee, Fisheries Development Council (Department of Fisheries, Vancouver, revised December 1961) (Hereafter "Summaries of Research").

35. "Summaries of Research," p. ix; Mighetto and Ebel report that fish passage facilities at U.S. Army Corps of Engineers facilities (exclusive of research) cost an estimated "$130 million, with an annual maintenance cost of $1 million in 1956." Mighetto and Ebel, *Saving the Salmon*, p. 104.

36. NAC, RG 89, Vol. 672, File 2558A, J. R. Brett, "Salmon Research and Hydro Power Development," Fisheries Research Board of Canada, Biological Station, Nanaimo, BC, July 15, 1956, emphasis in the original.

be surmounted by a thorough program of research directed at the fish first, from which the problems may then be resolved.

Making the same point in 1957, W. S. Hoar, a UBC zoologist, stated that it would be "impossible" to make sound predictions about the effects of water development without first establishing the "critical biological data."[37] Rather than be led by a problem-oriented agenda, by which fisheries scientists tried to solve any given dam problem when and if it arose, fisheries scientists argued for the necessity of a broader vision that would help to establish baseline knowledge from which specific decisions could be made.

Fundamental research remained nevertheless an imprecise goal. Different scientists believed that some problems required more or less attention. In Brett's detailed schema of future research, he identified three broad fields of concern: physiology, stress, and migration. Within these three fields, he identified twelve topical areas of investigation.[38] Unlike the opening emphasis on fundamental research, however, some of them focused on applied problems such as how to handle fish passage during the construction phase of a dam or how to entice salmon into bypasses and around dams. The line between fundamental and applied research blurred when specific topics were entertained. Other researchers, such as Peter Larkin and W. S. Hoar, suggested the priority of other kinds of research. Larkin, with a strong research background in limnology, argued for close attention to impacts of dams on lakes, water temperatures, and water-quality issues.[39] Hoar, a zoologist, drew his brief on the basis of his reading of problems in ecology, physiology, and fish behavior: He advocated a broad approach drawing from and linking each of these fields.[40] More than Brett, who had

37. UBC, Fraser River Hydro and Fisheries Research Project Papers, File 4, William S. Hoar, "Power Development and Fish Conservation on the Fraser," Technical Report – Biological, September 15, 1956, emphasis in the original.
38. These were delay, changes in water quality, diversion, blocked passage, passage upstream through reservoirs, final upstream migration, passage downstream through reservoirs, passage over dams, bypass entrance and transportation, condition of young salmon in tailraces, changes in water quality in terms of downstream migration, and problems during the construction phase.
39. Larkin, *Power Development*.
40. UBC, Fraser River Hydro and Fisheries Research Project Papers, File 4, William S. Hoar, "Power Development and Fish Conservation on the Fraser River" Technical Report – Biological, September 15, 1956.

applications in mind, Hoar had a vision of a wide field of investigation, lasting over at least three to four sockeye life cycles, or fifteen to twenty years. Fundamental research branched out into a variety of directions for Hoar: from long-term studies of the evolutionary shifts in salmon after the last Ice Age to close studies of salmon behavior. What went without saying from all of these students of the problem was that their ideas of fundamental fisheries science were nevertheless directed: toward the river, its tributaries, and lakes and away from the ocean. Fundamental science, in these readings, contained a spatial bias.

Apart from the fact that no single vision of research needs could organize fisheries science on the fish–power issue, institutional differences imposed different kinds of restraints, obligations, and possibilities on researchers. Brett's prospectus for fundamental research grew out of earlier studies conducted in the Fisheries Research Board labs and field studies; his perspective developed out of a particular context. Larkin's and Hoar's programs, on the other hand, arose from discussions surrounding the BC Electric research grant to UBC for fish–power research. Their positions were framed by the question put to the university by BC Electric President Dal Grauer: Could there be low-level dams on the Fraser and still be healthy fish runs?[41] They imagined their research programs in terms of the mandate of the university, as opposed to government research, and wondered how research contracts might limit academic freedom.[42] Fundamental research and the appropriate research directions accordingly had different meanings for both groups. Beyond these institutions, other active research agencies pursued applied programs, in keeping with their mandates: The Fish and Game Branch focused on lake productivity and limnology studies with a view to the sports fishery; the IPSFC investigated dam proposals on the Fraser on a case-by-case basis in order to contain projects that might infringe on the Pacific Salmon Convention; and the Fish Culture and Development Branch of the Department of Fisheries pursued mitigation work at existing and prospective dams

41. UBC, Fraser River Hydro and Fisheries Research Project Papers, File 2, E. S. Pretious et al., "Fish Protection and Power Development on the Fraser River" (UBC, February 1957). These authors quote Grauer's question to the researchers as part of their terms of reference.
42. Larkin, *Power Development*, p. 38.

and carried out research on the possibilities of artificial spawning channels and hatchery production. Thus the call for fundamental research was in some sense not only an attempt to organize and redirect research in progress but also a demand for an expansion in the capacity of fundamental research. In his review of fish–power research, W. S. Hoar wrote of the necessity of "funds commensurate with the magnitude of the proposed changes."[43]

Yet if not all fisheries research on the West Coast could feed into a single coherent fundamental research program, there was much work of value contained in the different institutional spheres and applied programs. Over the 1950s, research on fish–power topics and related matters blossomed in terms of research effort and the number of projects (see Table 4). When fisheries scientists compiled an index of their work – both fundamental and applied – at decade's end, they recorded a steady growth in projects. By the middle of the decade, researchers scrambled to keep up with the pace of the work and the range of new findings.

To shape this evolving program, scientists engaged in new forms of cooperation within and between institutions. Marked formally by interagency committees and joint research projects, cooperation also operated informally as scientists shared expertise, research findings, and future plans. Although a hallmark of much past fisheries science in BC, cooperation took a qualitative shift in scale and significance under the political pressures of the fish–power problem. Fisheries scientists operated on the belief that time was of the essence and that the pooling of expertise and research funds and the creation of a synergistic research environment provided the most direct route to a "solution."

The most concrete expression of this new cooperative impulse revealed both the desire to enhance cooperative research outcomes and the problems of political pressures. Formed in 1957 under the auspices of the federal Department of Fisheries, the research subcommittee of the Fisheries Development Council brought together representatives from all five of the major fisheries research institutions in BC: the federal department, the FRBC, the provincial Fish and Game Branch, the IPSFC, and the UBC Institute of

---

43. UBC, Fraser River Hydro and Fisheries Research Project Papers, File 4, William S. Hoar, "Power Development and Fish Conservation on the Fraser River," Technical Report – Biological, September 15, 1956.

Table 4. *The Growth of Fish–Power Fisheries Research Measured by Project Starts*[a][44]

Research Project Start Dates	Total Projects of All Fisheries Research Agencies
Before 1950	34
1951–3	31
1954–7	68
1958–60	51

[a] NB: Because the table measures research starts, it hides the number of projects extending over more than one time period. It should be noted that, on an aggregate level, the number of total projects in operation grew over time. Fifty-nine projects extended into the final research period that had begun before that time.

Source: "Summaries of Research."

Fisheries. It aimed to provide a forum for the sharing of research findings, the exchange of ideas, and the creation of joint projects. The connection of the subcommittee to the broader council, however, suggested the second aim of the federal department in convening the committee: to advise the council's industrial and union representatives on the course of fish–power research, provide technical background positions on current disputes, and engage with the fisheries interest in planning a defense to the power threat. One of the subcommittee's practical achievements spoke to both its research and political mandates: a summary document, created in the late 1950s that provided briefs on all fish–power research projects or related projects carried out in BC since the late 1930s. This document proved useful to active researchers as a way of linking projects and keeping tabs on a burgeoning field, but it also gave the Fisheries Department and industrial concerns a sense of the direction of research effort and some of the basic findings. For purposes of analysis, it also contained a good deal of useful, if uneven evidence, from which to draw conclusions about the organization and course of research in these years.[45]

44. Based on information contained in the "Summaries of Research."
45. NAC, RG 23, Vol. 1230, File 726-11-14, part 2, A. J. Whitmore, director, Pacific area to deputy minister of fisheries, December 10, 1958; "Summaries of Research." In interpreting this document I have followed the committee's definition of research relevant to

Table 5. *Research Institutions and Cooperative Projects*[46]

Primary Investigator Institution	Research Projects	Cooperative Projects	Cooperative Projects as % of Total
Department of Fisheries, Canada	76	41	54
BC Fish and Game Branch	24	10	42
FRBC	20	11	55
IPSFC	51	22	43
UBC	14	12	86

Source: "Summaries of Research."

In summarizing their work, fisheries scientists found it note-worthy to record the extent of cooperation among institutions in the course of any single investigation. Cooperation was both a practice and a matter of observation. Of the 185 projects summarized by the research subcommittee, more than half of them were cooperative in design. From the listings of the primary investigator institution and their research partners, it is possible both to weigh the importance of cooperative projects to different institutions and identify the range and importance of research linkages.

Table 5 suggests the wide extent of cooperative projects across the institutional landscape, with the strongest cooperative emphasis appearing in the work of the UBC Institute of Fisheries and the federal Department of Fisheries. In these cases, access to research facilities and field locations proved important for university researchers, and matters of jurisdiction and technical expertise drove the federal department's cooperative needs. The fact that the majority of research projects on fish and power were conducted within the Fraser Basin, for example, drew the federal department into cross-jurisdictional relationships with provincial authorities and the IPSFC.[47]

the fish-power problem. Some of the research listings are incomplete or do not follow a consistent pattern (for example, in terms of project cost reporting). I have attempted to account for such difficulties as far as possible.
46. Based on information contained in the "Summaries of Research."
47. Based on information contained in the "Summaries of Research."

Turning to the emphasis of cooperation between institutions, however, one finds stronger linkages between certain key groups. The Department of Fisheries, with the greatest number of cooperative projects, recorded fifteen separate cooperating agencies, from the U.S. Fish and Wildlife Service to the BCPC to the UBC Institute of Fisheries. Its greatest links were with the provincial Fish and Game Branch (eighteen cooperative projects) and the IPSFC (twelve cooperative projects). For other agencies, the Department of Fisheries proved to be the most important cooperating partner: It ranked as the lead cooperator for the Fish and Game Branch, the FRBC, and the IPSFC. Only the Fisheries Institute at UBC did not maintain strong research cooperation linkages with the federal department; its major cooperating agency was the National Research Council, keeping with its academic orientation. Thus in parallel with the important coordinating function carried out by the federal department through the research subcommittee of the Fisheries Development Council, the department also acted to connect different institutions on a practical, research project level.[48]

If cooperation was crucial to the expansion in research and the conduct of a more unified program, it did not solely define the institutional nature of research in this period. Over the 1950s, there was a rough fifty–fifty balance maintained by institutions between cooperative and individual projects. Cooperation occurred at what might be described as a middling band of research funding. Projects with budgets under $10,000 recorded a majority of individual projects, and for those over $50,000 the balance was almost equal. Between those posts, however, the majority operated with a cooperative element. Smaller projects, such as individual stream surveys, for example, did not require cooperation. The biggest projects, on the other hand, tended to have some mitigation aspect, specific to a particular agency, which may not have been of great import for research, but generally had high capital costs. The major research effort, in the middle band, however, was squarely cooperative. It tended to involve more personnel, extended over longer periods, and involved two or more cooperating agencies. Thus, although individual institutional projects continued and increased over the

48. Based on information contained in the "Summaries of Research."

period, the cooperative projects proved of particular importance to expanding research horizons.[49]

Cooperation meant different things in different contexts. In the case of the UBC-BC Electric research grant, for example, there was a strong emphasis on blending the work of engineers with fisheries biologists so that the different insights of both might be brought to bear on the problem and help to sharpen the weak sides of both disciplinary approaches. Yet, despite this attempt to create interdisciplinary cooperation, little effort was made by the UBC research team to link their projects with other fisheries agencies. In part this was conditioned by the nature of the grant donor. BC Electric was not interested in hiring or funding government scientists or science, but looked to the university as an impartial research institution, elevated from the politics of fish vs. power disputes in which government agencies were presumably embroiled. In a preliminary planning document for the project, Peter Larkin stated that the university was looked on as a disinterested observer; its pronouncements on the fish–power issue were neither invited nor desired.[50] However, the university did play a cooperative function that is not observable in the figures of cooperative projects by playing host to a number of research conferences on the fish–power issue. One such conference, funded by a special grant from H. R. MacMillan, included scientists from the federal and provincial fisheries agencies.[51] In this case, the donor hoped to integrate results and expand the discussion. Cooperation could thus operate at a variety of levels and with different emphases, depending on the nature of the political and institutional contexts.

Cooperation had a transnational element, but not as strong as one might have expected. Some general sharing of research results occurred between scientists and agencies in both Canada and the United States, and transnational cooperation remained a cornerstone of the IPSFC mandate and program. Joint research projects also proceeded when sharing personnel or facilities made sense. One project, for example, pooled the skills of scientists from the IPSFC, Department of Fisheries, and the Washington State

49. Based on information contained in the "Summaries of Research."
50. Larkin, *Power Development*, p. 38.
51. The conference was held April 29 and 30, 1957, and will be subsequently discussed.

Department of Fisheries to investigate the survival rates of down-stream migrants through dam turbines at a suitable American facility, Baker dam in Washington State.[52] Yet beyond such examples, cooperative research with American agencies did not become a major priority, despite the similar concerns and problems of American researchers. In part, Canadian researchers hoped not to duplicate American projects needlessly and so envisioned their work in conscious distinction from American precedents. More critically, some Canadian scientists believed that American research remained mired in an overly applied phase. In his review of fish–power research, for example, J. R. Brett criticized American research on the Columbia as overwhelmingly focused on local technical problems relating to particular dam structures rather than on general questions of fish physiology and ecology.[53] Canadian research displayed a more fundamental element, in his view. Although this conceit is questionable, given the scale of applied work in Canada in the same period, it is also true that much American research funding remained tied to mitigation work at the Columbia dams in the 1950s for the very pressing reason that salmon numbers were falling and scientists wanted to know why and how the situation could be improved. On the other hand, the Canadian problem was relatively less pressing and so fundamental research could be engaged in a different way. It should also be remembered that Canadian and American research grew out of different political and institutional contexts. Joseph Taylor argues, for example, that artificial propagation research dominated American research over the first half of the twentieth century, whereas it played almost no role in Canada following the cancellation of federal hatcheries in 1935. American research remained tied to hatcheries and mitigation work while Canadian research followed a more flexible course.[54] Nevertheless, these two research communities did maintain research ties within and without the IPSFC and profited from close contacts at conferences and in correspondence. The conduct of fish–power research proved to be much more focused on the integration of national,

52. Hamilton and Andrew, "An Investigation of the Effect of Baker Dam."
53. NAC, RG 89, Box 672, File 2558A, J. R. Brett, "Salmon Research and Hydro Power Development," Fisheries Research Board of Canada, Biological Station, Nanaimo, BC, July 15, 1956, foreword.
54. Taylor, "The Political Economy of Fishery Science."

provincial, and international institutions, rather than creating expanded transnational links with U.S. or state counterparts.

Answers and Expertise

In the early 1950s, fisheries scientists had called for more research to solve the fish–power problem. They had attempted to manage expectations and explain the extent of their ignorance. By the closing years of the decade, however, granting agencies, governments, public commentators, and dam developers called for answers. The research money had been spent; now where were the results and lessons? Fisheries scientists understood the demands for answers, and they responded in kind with conferences, special synthetic reports-on-progress, academic papers, and commentaries. In general, it might be said that fisheries scientists offered no panaceas and spoke cautiously about the capacity of technical contrivances to overcome the fish–power dilemma. There would be no miraculous discovery as at Hells Gate and no immodest predictions of the capacity of fishways to domesticate dams. Instead, scientists came to a fuller appreciation of the limits to remaking salmon under new environmental conditions. This was hardly the ringing response that promoters and power companies had hoped for, and various challenges to scientific authority and knowledge occurred. In response, fisheries agencies and institutions confronted critics and defended their research.

A conference held at UBC in 1957 to explore the problems of fish and power demonstrated the reticence of fisheries scientists to declare a solution and their desire to underline the extent of the difficulties. Introduced by the distinguished UBC fisheries biologist Wilber Clemens, the conference proceeded with a sense of gravity about the role of fisheries biology in public policy. Clemens expressed the concern thus[55]:

... [T]he biologist is asked whether or not salmon may be passed over or around dams; if they can, how; and if not, why not; and in the latter case

55. W. A. Clemens, "The Fraser River Salmon in Relation to Potential Power Development," in Larkin, *The Investigation of Fish–Power Problems.* pp. 3–10.

whether there is any method of maintaining the salmon stocks and having hydro-electric dams. Upon his answers hang very important engineering, economic and social consequences. Upon his answers rest decisions that will affect the lives of very many people for very many generations to come.

Evaluating the many dam proposals before the public, Clemens rattled through the solutions aired in the press: trucking salmon around dams, massive fish-passing devices, and lift machines. He wondered how any such plans could cope with the massive scope of passing up to tens of thousands of fish per hour in main-stem dams during peak migration times, or whether the imaginable changes in water temperature and quality, current, and diversion would impose too great a burden on salmon. He did not say that these problems could never be solved, but his menu list of difficulties suggested his current pessimism.

In subsequent lectures speakers addressed the problems awaiting solution. E. H. Vernon, a biologist with the BC Game Commission, analyzed the impacts on lakes of reservoir flooding. W. R. Hourston, chief biologist with the Department of Fisheries, Pacific Area, followed Clemens in a general assessment of the many problems attending changed water and passage conditions. Stream ecology came under scrutiny from Ferris Neave of the FRBC, and problems of lactic acid buildup in delayed fish from Edgar Black, a UBC physiologist. J. R. Brett of the FRBC outlined a general theory of stress effects, and G. Collins of the U.S. Fish and Wildlife Service described the course of experiments in existing Columbia River dam fishways. F. E. Fry of the Ontario Fisheries Research Laboratory discussed how to measure performance in fish, and William Hoar of UBC, how to analyze behavior.[56] Here then was a summary of much of the ongoing work of a variety of different agencies, concerned with related, but different, scientific and regulatory

56. The following chapters are found in Larkin, *The Investigation of Fish–Power Problem*: E. H. Vernon, "Power Development and Lakes in British Columbia," pp. 11–14; W. R. Hourston, "Power Development and Anadromous Fish in British Columbia," pp. 15–24; Ferris Neave, "Stream Ecology and Production of Anadromous Fish," pp. 43–8; Edgar C. Black, "Energy Stores and Metabolism in Relation to Muscular Activity in Fish," pp. 51–67; J. R. Brett, "Implications and Assessments of Environmental Stress," pp. 69–83; G. Collins, "The Measurement of Performance of Salmon in Fishways," pp. 85–91; F. E. Fry, "Approaches to the Measurement of Performance in Fish," pp. 93–7; William S. Hoar, "The Analysis of Behaviour in Fish," pp. 99–111.

problems. These were reports-on-progress, but also statements of concern about the possibility of answering the questions posed. D. S. Rawson, a fisheries biologist at the University of Saskatchewan, introduced a self-critical reference in the course of a general discussion of limnology that revealed the level of doubt about answering the problems, or of even asking the right questions[57]:

Another obstacle to the solution of our problems is that we are not too much concerned with the primary biological production but rather with the production of fish; and not even the total production of fish, but in the production of certain kinds of desirable fish. Biologically speaking, the fisherman takes a very narrow and prejudiced view of productivity. Thus, if we cannot predict productivity in a natural lake, should we try it in a disturbed lake, where in any case, we are only interested in a very special kind of production?

Rawson's question might be said to summarize well the concerns held by these biologists, not only about the ability to maintain fish runs in the event of dam building, but also about their capacity to understand those changes in their full significance. They offered their best answers to the pressing public demands, outlined initially by Clemens. What they could not do was offer final answers or imagine their possibility in the short term.

During the 1950s supposed advances in fisheries science commanded newspaper copy, and this conference was no exception. In the midst of the event and for days afterwards, metropolitan newspaper articles and editorials reported the scientists' findings and tried to make sense of them. Editorialists found much to suggest optimism, in contrast to the skepticism evident in the text of the lectures. Indeed most newspapers reported that a solution to the fish–power problem loomed. As the *News Herald* enthused, "Fish Doctors Work Out Lines to Put Salmon in the Pink."[58] The lecture presented by Ferris Neave of the FRBC attracted considerable attention and dominated coverage to the exclusion of the keynote address and other major papers. Its subject – stream ecology – was not the reason, but rather a casual remark Neave offered to the effect that dams as dams are not necessarily harmful to fish, but can

57. D. S. Rawson, "Indices of Lake Productivity and Their Significance in Predicting Conditions in Reservoirs and Lakes with Disturbed Water Levels," pp. 27–42 in Ibid.
58. BCER CF, *Vancouver News Herald*, May 1, 1957.

in some instances help to protect spawning grounds from floods. When pressed on the point during questioning, Neave stated that not all dams performed this function, certainly not hydro dams, but he playfully imagined that dams could be made with a view to regulating flow to enhance salmon production.[59] Transmogrified into newspaper print, however, this idea became a statement that fish and dams could coexist. Better still: Dams could *benefit* fish. Headlines extolled "Beneficial Dams" and remarked that "Proper Dams Help Fish."[60] The only articles that engaged scientists' skepticism used their views as an argument for expanded research to solve the problem and quickly. The *Sun* reported "the frank confession of man's ignorance about the ways of Pacific salmon," and concluded that the conference "raises the need for salmon research almost in the crisis category."[61] What research had been done and what its implications were for the fish–power problem remained matters of less comment.

The range of newspaper coverage in this instance – its emphasis and optimistic selectivity – found parallels in the general public discussion of fish–power research. Like the focus on the beneficial effects of dams, numerous optimists suggested that, fisheries scientists to the contrary, the fish–power problem was virtually solved. One of the most vocal proponents of this view was Val Gwyther, a Vancouver consulting engineer, member of the Fraser River Multiple Use Committee, and self-educated expert on the fish–power issue. On a number of occasions during the late 1950s, Gwyther attracted publicity for his heterodox views on solutions to the fish–power problem. He was described in private by one fisheries lobbyist as "either a mental case or a confirmed 'time-waster'" for the simple reason that his views ran utterly contrary to the skepticism of fisheries scientists and that he received considerable attention for those views.[62] Gwyther's central contention was that an engineered river would prove more beneficial to salmon production

59. Neave, "Stream Ecology."
60. BCER CF, "Proper Dams Help Fish, Says Expert," *Vancouver Sun*, April 30, 1957; "Power Versus Fish Problem Solved by 'Beneficial Dams,'" *Vancouver News Herald*, April 30, 1957; "Salmon Aren't Smart But They Puzzle Us," *Vancouver Sun*, May 1, 1957.
61. BCER CF, "Salmon Research Urgent," *Vancouver Sun*, May 2, 1957.
62. UBC, Fisheries Association of BC Papers, File 45-34, unsigned statement dated June 8, 1958, concerning Gwyther's paper, "Some Facts – Fish and Other Resources."

than the "natural river."[63] He had read and mastered much recent American research on fish-passage techniques and presented it in digested form for Canadian engineers; he had made calculations of the monetary value of the Fraser River per acre-foot according to different industrial uses; he had also become much impressed by the artificial propagation methods employed by U.S. fisheries regulators to perpetuate runs lost to dams.[64] Dams and fish could live together, he concluded, and salmon runs could be expanded once natural spawning, "with its random and inefficient yearly production," was improved by the scientists' helping hand.[65] In newspaper articles, and the engineering press, Gwyther's ideas received wide public notice.[66]

If the attention to Gwyther's ideas suggests the eagerness of the press to latch onto notes of optimism, the counterreaction to his ideas offered by fisheries scientists speaks to the anxieties of established scientific authority when dominant conclusions received widely reported criticism and opposition. To stem the positive press reports of Gwyther's ideas, fisheries scientists and officials systematically worked to destroy his reputation. American scientists knowledgeable about artificial propagation wrote public declarations condemning Gwyther's interpretation of their work and asked him privately to cease his declarations; Tom Reid of the IPSFC publicly decried Gwyther's work as "trash"; and fisheries officials privately contacted Gwyther, criticizing his views.[67] To all of this Gwyther could reply only that fisheries scientists did not have privileged access to the truth. "I suggest to you," he wrote to Dr. A. L. Pritchard of the federal Department of Fisheries, "that developments of any kind can only be evolved with a broad logical

63. Val Gwyther, "Watershed Resource Value."
64. UBC, Fisheries Association of BC Papers, File 45-34, Val Gwyther, "Multiple Development of the Fraser River Basin: The Solution to the Conflict of Fish and Power," *BC Professional Engineer* 9(10) (1958): 13–19; and "Some Facts – Fish and Other Resources."
65. BCER CF, "Fraser Dams Could Prove Salmon Boon," *Vancouver Sun*, January 8, 1959.
66. For example, BCER CF, "Fish and Power Can Co-Exist for Benefit of BC Economy," *Columbian*, October 24, 1958; "Fraser Dams Could Prove Salmon Boon," *Vancouver Sun*, January 8, 1959.
67. BCER CF, "Dams Won't Aid Salmon, Official Says," *Province*, January 10, 1959; "Senator Hits Power Dam Advocates," *Vancouver Sun*, January 22, 1959; "Engineer's Fish Claims 'Incorrect,'" *Vancouver Sun*, March 19, 1959.

approach to the problem; an approach that must take in and analyze all data that is available from any standpoint and resolve impartial conclusions."[68] In the most substantial critique, C. H. Clay, a federal fisheries engineer, and Peter Larkin of UBC coauthored an article later distributed to MPs, MLAs, the press, and radio, with a bold title intended to stop Gwyther's ideas in their tracks: "Artificial Propagation is NOT the Answer."[69] The concern of fisheries interests not only to attack Gwyther's ideas, but also to label him as a nonexpert without the capacity to judge, is suggested in the authorial background to this critique. C. H. Clay of the federal Department of Fisheries wrote the piece independently. Members of the Fisheries Development Council, however, believed that the power of the rebuke would be strengthened if another respected scientist, beyond the federal department, appeared as the paper's coauthor. Peter Larkin was approached as a result and agreed.[70] This addition, as well as Larkin's willingness to have his name used in this way, suggests how "cooperation" could mean working hard to exclude others as well as joining together. It also demonstrated the direct links between supposedly disinterested scientists and the unapologetically interested fisheries lobby.

Gwyther suffered considerable public shaming at the hands of these fisheries scientists, but he shared company with others. Harry Warren, long one of the most vocal, and, from a fisheries point of view, reckless dam promoters, faced considerable and sustained public and private attacks on his integrity. Labeled publicly as "a geologist who knows nothing about fish," in Tom Reid's dismissive phrase, Warren also suffered attacks within the university.[71] In 1958 after Warren delivered a series of public speeches on the Moran dam, federal Fisheries Minister James Sinclair warned UBC

68. NAC, RG 23, Vol. 1225, File 726-11-5[14], Gwyther to A. L. Pritchard, December 2, 1958.
69. BCER CF, "Engineer's Fish Claims Incorrect," *Vancouver Sun*, March 19, 1959; "Dams Won't Aid Salmon, Official Says," *Province*, January 10, 1959; "Senator Hits Power Dam Advocates," *Vancouver Sun*, January 22, 1959; UBC, Fisheries Association of BC Papers, File 45-34, Milo E. Moore, director of fisheries, State of Washington, to Val Gwyther, February 13, 1959 (copy); Clay and Larkin, "Artificial Propagation is NOT the Answer."
70. NAC, RG 23, Box 1225, File 726-11-5, part 15, A. J. Whitmore to deputy minister, March 4, 1959. Whitmore discusses the meeting and decision in this letter.
71. BCER CF, "Senator Hits Power Dam Advocates," *Vancouver Sun*, January 22, 1959.

President Norman Mackenzie of looming public criticism of the university if he did not rein in Warren[72]:

I write this letter to advise you that while the spokesmen of the fishing industry have so far been restrained in commenting on the foolish utterances of Professor Warren in this field which is foreign to his training, restrained mainly out of respect for the University, you cannot expect this restraint to be maintained if the Professor continues to use his standing as a Professor of a great University to advance the interests of American promoters.

In a parallel episode, engineering professors Eugene Ruus and J. F. Muir came under attack from fisheries defenders for their views and supposed links to power interests. Both professors were members of the Fraser River Multiple Use Committee and authors of an optimistic review of engineering research on fish–power matters that proposed, among other things, that hydro development in the Fraser Canyon would solve the Hells Gate problem by flooding it under a reservoir.[73] Tom Reid, chair of the IPSFC, claimed to the press that the two engineers were apologists for BC Electric because the company funded their research. Although the engineers received research funds from the UBC president and the Fraser River Multiple Use Committee, they in fact had no connection to BC Electric.[74] It is difficult to imagine that Tom Reid, with close connections at UBC and in fisheries science more generally, did not know this point, but was attempting instead to discredit opponents however he could. In a related and extraordinary act, Loyd Royal, scientific director of the IPSFC, wrote to UBC President Norman Mackenzie and, without naming Ruus and Muir, called for the president to intervene and redirect the research programs of certain

72. UBC, Fisheries Association of BC Papers, File 75-1, James Sinclair, minister of fisheries to N. A. M. Mackenzie, president, UBC, February 28, 1957 (copy).
73. BCER CF, "Hydro Dam Scheme at Spuzzum Won't Impress Fishery Experts," *Columbian*, June 3, 1961; "BC Engineers Solve Fish–Power Problem," *Victoria Colonist*, June 2, 1961; Muir and Ruus, "Engineering Research on the Fish Power Problem."
74. BCER CF, "2 Professors Deny Influence of BCE," *Vancouver Sun*, June 12, 1961; "Reid Charges BCE Seeks Fraser Dam," *Vancouver Sun*, June 9, 1961. Fisheries biologists also made public statements questioning their research: BCER CF, "Biologists Say Fishway Proposals are Fanciful," *Province*, June 6, 1961. A grant of $500 to support Muir's research and his acceptance of said amount is recorded in the minutes of the committee: UBC, Fisheries Association of BC Papers, File 45-34, "Minutes of the Seventh Meeting of the Fraser-River Multiple Use Committee, June 3, 1958."

professors engaged in areas beyond their area of expertise. "It is my personal request," wrote Royal, "that you give serious consideration to eliminating the current trend of activity by these men and give personal support to an expansion of activity by them limited to their respective fields of endeavor."[75] Apart from searching for a solution to the fish–power problem, fisheries scientists and officials made considerable efforts to reinforce their authority and denounce individuals who offered contrary views, whether they held scientific credentials or not.

A different challenge to scientific authority emerged from the many unsolicited solutions proffered by amateur tinkerers and inventors to the fish–power problem in the 1950s. Originating from outside any formal educational or professional setting, various designs for miraculous fish passage devices made their way into newspaper pages and landed on the desks of bemused scientists engaged in fish–power research. The total number of designs of this kind is unknown, but it was sufficient to cause complaint from the scientists who were asked to assess them and led former federal Fisheries Minister James Sinclair to remark in 1958, "Almost every inventor in Canada seems to have had a crack at trying to devise some way to pass fish around high level dams."[76] The problem of passing fish over dams proved irresistible to a wide range of backyard inventors, who included, for example, Cecil Wilkinson of Victoria, an estimator for a moving firm, who developed a massive waterwheel with "six pivoting blades, which swing open on one side and remain rigid on the other as the wheel turns."[77] Such offers of assistance were entirely innocuous, genuine, sometimes ingenious, and frequently bizarre. But the manner in which such solutions were judged by scientists and handled by the press spoke to broader issues of defining expertise.

One of the most widely publicized designs, the so-called Devlin fishway invented by Powell River machine superintendent, A. G. Devlin, quickly became a minor cause célèbre for small-town dam

75. NAC, RG 23, Box 1224, File 726-11-5, part 11, Loyd Royal, director, IPSFC, to Dr. Mackenzie, president, UBC, December 6, 1956.
76. BCER CF, "Develop Power Resources on Non-Salmon Rivers," *West Coast Advocate*, April 23, 1959.
77. BCER CF, Jack Fry, "Hydro 'Water Wheel' Might Save Salmon," *Victoria Colonist*, March 13, 1960.

promoters critical of scientific studies of salmon conservation. Following a demonstration of the device at UBC in 1959 that led scientists to judge it wanting, numerous local newspapers carried articles on the fishway and its inventor, complete with a crude line drawing showing a model dam with angular passages, full of arrows and small penned salmon swimming through effortlessly. Harry Taylor, writing in the *Ashcroft Journal*, noted the unhelpful response of UBC scientists to this remarkable design ("All the fisheries experts can say is 'it won't work'"), despite positive reactions from some unnamed engineers. He ended his article by asking "what more do the experts demand?"[78] Similar articles followed in newspapers as diverse as the *Whalley Herald*, the *Powell River News*, and the socialist paper *West Coast Advocate*.[79] After some of these reports, IPSFC representatives penned replies that spoke to some of the difficulties of fish passage.[80] In the *West Coast Advocate*, Tom Parkin of the UFAWU wasted no time in explaining the Devlin design as an ideological tool of the power establishment: "Despite the claims of self-styled experts who speak for the power people, trained fishery biologists and engineers both in the U.S. and Canada claim there is no solution to the problem even in the foreseeable future."[81] This condemnation accepted the authority of science as impartial and ideologically neutral and categorized lay inventors as pawns of power. Although the Devlin design led nowhere, the debates over its efficiency suggest the frustration of certain interests with the seeming intransigence of "experts" to commonsense designs and the importance to fisheries interests of ensuring that those designs were understood as amateurish.

Despite much criticism from the likes of Harry Taylor, UBC scientists involved in fish–power research showed remarkable patience in taking time out from their primary responsibilities to play host to a variety of amateur demonstrations.[82] On one such occasion, UBC

78. BCER CF, Harry Taylor, "Hydro Power on the Fraser or Salmon or Both?" *Ashcroft Journal*, March 26, 1959.
79. BCER CF, Harry Taylor, "Hydro Power on the Fraser or Salmon or Both?" *Whalley Herald*, March 12, 1959; *Powell River News*, January 22, 1959; "Develop Power Resources on Non-Salmon Rivers," *West Coast Advocate*, April 23, 1959.
80. BCER CF, Tom Reid, "The Case for Fraser Salmon," *Whalley Herald*, March 12, 1959.
81. BCER CF, "Develop Power Resources on Non-Salmon Rivers," *West Coast Advocate*, April 23, 1959.
82. UBC, Fisheries Association of BC Papers, File 45-33, "The Fraser River Hydro and Fisheries Research Project Progress Report" (Vancouver: UBC, July 21, 1958). This

engineering professor Edward Pretious hosted three visits from one Albert E. Dane, a Canadian World War I veteran and California resident, who had developed theories about fish passage. "My background," Dane explained in a preliminary letter, "is non-academic, yet for many years I have been engaged in what may be termed a hobby, having primary significance wherever the term 'fluid' has meaning. Apart from other developments, I have applied my understanding to the problem of fish passing 'at any desired angle and under varying conditions of flow etc.'"[83] Dane's letter first reached General McNaughton, following on the much-publicized proposal to divert the Columbia into the Fraser. It then passed to the Department of Fisheries and from there to Edward Pretious at UBC, who said he believed in the importance of remaining open to ideas, no matter their source.[84] But Dane's presentations to Pretious did not impress. Reporting on the meetings to federal fisheries officials, Pretious gave this explanation[85]:

Mr. Dane was anxious to impress me at first with the fact that he is not particularly concerned with fish as such, but only in the behaviouristic responses of all living creatures to the fluid medium in which they live, whether it be air or water. This rather general treatment of animal creation made me cautious and I subsequently discovered that his knowledge concerning the characteristics of anadromous salmon was rather hazy. I say this because, recently I have had the rather onerous task of corresponding with members of the public at large, who feel very confident that they can solve the fish-passage problems of salmon migration much better than all the conservationists, engineers and biologists who have spent so much time, money and talent on these problems. Maybe they can, but until their ideas are put to the test the world will never know, because unfortunately no one is willing to gamble on these ideas. Furthermore, these well-meaning individuals put me in mind of people who are quite prepared to perform an intricate and delicate surgical operation without even the benefit of an introductory course in first aid. They personally do not stand to lose anything which makes them very confident.

reports states that both Peter Larkin and Edward Pretious fielded numerous public "solutions" to the fish vs. power problem.
83. NAC, RG 23, Vol. 1229, File 726-11-10[2], Albert E. Dane to General A. G. L. McNaughton, Canadian chairman, IJC.
84. NAC, RG 23, Vol. 1229, File 726-11-10[2], Edward Pretious to C. H. Clay, Department of Fisheries, October 29, 1956.
85. Ibid.

Monitoring the ideas of amateur inventors, giving them a hearing and explaining their faults drew both from a sincere willingness to accommodate ideas from unexpected places, but also a concern to nip naïve suggestions in the bud. The expert bore responsibility and obligations, Pretious argued; the amateur did not.

Despite the many such challenges presented to fisheries scientists and their authority, the opposite condition of an uncritical acceptance of fisheries scientific expertise imposed a different set of difficulties. During the late 1950s, after an initial stage of fisheries research suggested the sheer complexity of the problem, numerous interests, from widely different political perspectives, joined in calling for an expansion in researching funding. It seemed to some that funding for fish–power research had now reached a "crisis stage," as the *Sun* declared following the UBC fish–power research conference in 1957. "Unless Canada is prepared to accelerate its research for a solution to the growing problem of fish and their peaceful co-existence with power," stated BC Electric Executive Director Harold Merilees, sounding like a cold warrior, "it may be too late – BC's economy may demand commencement of hydro development on the Fraser."[86] Harry Warren supported more research, as did John Deutsch, chair of the UBC economics department; the *Western Fisheries Magazine*, the Fisheries Association of BC, and former federal Fisheries Deputy Minister Donald Finn all made it known that more research funding would go a long way to overcoming the current difficulty.[87] Gordon Shrum, chair of the BCPC and a public supporter of developing the Fraser, proposed a government-led $10 million research program in 1958 to put the matter of fish and power to rest.[88] With a "joint federal–provincial research program," ran an editorial in the *Sun* the same year, a solution to fish and power might be possible.[89]

86. BCER CF, "Power Expert Urges Fraser River Research," *Cariboo Observer*, March 18, 1959.
87. BCER CF, "Fish vs. Power Case Depends on Research," *Province*, June 4, 1958; "It's a Matter of Fish AND Power for BC, Economists Tell Panel," *Province*, March 2, 1957; "More Work Urged on Fish Study," *Vancouver News Herald*, January 29, 1957 (this article reports the *Western Fisheries* position); "Still Shy on Research," *Vancouver Sun*, February 17, 1958; "Research Answer to Fish vs. Power," *Vancouver Sun*, February 14, 1958.
88. BCER CF, "No Solution Yet," *Victoria Times*, April 30, 1958; "A Word from the Wise," *Vancouver Sun*, April 29, 1958.
89. BCER CF, "Real Threat to Fraser Fish Lies in Politicians' Blindness," *Vancouver Sun*, March 25, 1958.

The benefits to fisheries scientists of this widespread support for an expanded research program were obvious enough; what worried them was that it operated on the assumption that a solution *was* possible. For despite the seeming interest of power supporters in developing fisheries research, a strong current of belief existed that what was needed was an end to excuses and an expansion in large scale and effective research to force a solution. When Loyd Royal made the public statement in 1958 that a solution to the fish–power problem was still far in the future, for example, Gordon Shrum asked in a letter to the editor of the *Province* why fisheries scientists accepted such generous research funds when they seemed incapable of imagining a way ahead? "Fortunately," he asserted, "all scientists and engineers do not approach this problem with the same pessimistic and defeatist attitude."[90] His proposal for a major research program would, to the contrary, develop a solution within five years, a veritable fish–power "crash program."[91] Shrum, stated a *Province* editorial, held the "scientist's conviction that science can do almost anything."[92] Fisheries officials and scientists thus had to treat such support for research with caution and without accepting the responsibilities for instant results that power supporters wished to see. "Hydro authorities," said Tom Reid, chairman of the IPSFC, in one attempt at deflection, "are trying to create the impression that a solution of the fish–power problem is just a matter of probabilities, and that if enough people are given enough money, the problems will be solved in no time at all."[93] But what was needed, he and others repeated, was time and a greater appreciation that, as fisheries scientist R. N. Gordon put it, not all problems can "be solved by engineering principles alone."[94] In a major synthetic report of fish–power research published in 1960, F. J. Andrews and G. H. Geen wrote, "Extensive basic and applied research in salmon biology and fish-power problems is now being undertaken but there is no justification for expecting early solutions to all of the

90. BCER CF, letter to editor from Gordon Shrum, "It Takes a Smart Salmon to Fool a Scientist," *Province*, May 8, 1958; Royal's statement is contained in "Science Can't Fool Salmon," *Province*, May 2, 1958.
91. BCER CF, "Shrum Urges 'Crash' Research Program on Fish–Power Issue," *Province*, April 28, 1958.
92. BCER CF, "Science Can't Fool a Salmon," *Province*, May 2, 1958.
93. BCER CF, "No Easy Solution to Fish–Power Problem," *Columbian*, December 14, 1959.
94. Gordon, "Fisheries Problems Associated with Hydroelectric Development," p. 37.

particularly complex Fraser River fish–power problems."[95] Against what he viewed as a naïve techno-optimism, Peter Larkin put his own skepticism bluntly in 1960: "Anyone who believes that a pat universal solution to fish–power problems is around the corner is living in a fool's paradise."[96]

It was not until 1971, when the BC Energy Board, under the chairmanship of Gordon Shrum, reopened the possibility of a dam at Moran that fish and power again dominated the politics of the Fraser River and drew fisheries science into the orbit of provincial power policy. Yet on this later occasion, not only had the politics of power shifted with the rise of a new environmental movement in BC, but also the knowledge of the biology of salmon and the environmental consequences of dams had changed in significant ways. Now the problem of fish and power on the Fraser was framed with the background experience of large dams worldwide. Scientists traveled to Egypt to study the effects of the Aswan dam on the Nile to gain a sense of what might be the result of a dam at Moran on the Strait of Georgia. More directly relevant, the experience of fish and dams on the Columbia River inspired less confidence than formerly. Fisheries scientists learned after the 1950s that dams change fish, not just delay or block them. Reservoirs on the Columbia, for example, altered underwater gas conditions, producing imbalances in fish nitrogen consumption with the effect of producing large bubbles in fish bodies, which, when ruptured, killed fish. During the Moran investigations in 1972, Peter Larkin insisted that BC had the strongest research specialty in the world in problems of fish and power, that no solution could be easily imagined, that fish and dams could not simply coexist.[97]

Conclusion

The authority of fisheries scientists, notions of their expertise, and perceptions of their ability to transform nature rose, faced

95. Andrew and Geen, "Sockeye and Pink Salmon Production," abstract.
96. BCER CF, "Long-Term Plan Urged on Fish-Power Snag," *Province*, September 27, 1960.
97. BCARS, GR 442, BC Energy Board, Box 26, File 1, Larkin to Hugh Keenleyside, May 29, 1972.

challenges, and was vigorously defended over the 1950s. The problem of authority proved critical in the politics of fish and power. Who could say how salmon might react to environmental change, who had privileged access to the biology of salmon, and who could judge the right from the wrong had an impact on the public discourse of this environmental debate. In the early 1950s fisheries scientists gained widespread praise and trust for past successes. When initial results of investigation into the fish–power problem demonstrated only a more complex sense of limits, however, power promoters, inventors, amateurs, and enthusiasts sought to overcome a perceived pessimism and press the boundaries of established knowledge. These efforts resulted in few solutions, but in several contests of both minor and more general importance on the grounds of expertise and authority. Fisheries scientists defended their reputation and criticized those of apparently unqualified challengers as a means of maintaining a privileged and coherent voice in public discussion.

The insistence of fisheries scientists on denying their capacity to know, while at the same time questioning the knowledge claims of others, points to the politics of their studied claims of ignorance: No answer *was* an answer. While stating that science had no clear solutions, scientists also made clear that should politicians or developers wish to proceed, they did so with no scientific legitimacy or apology. They would have to admit frankly their willingness to risk destroying fish runs. Only the more flamboyant and marginal power promoters such as Harry Warren took this tack. Dominant hydro interests like BC Electric attempted to resolve the problem by funding science in search of a solution. Following the undecided reports on this research, and in the late 1950s the rising possibility of development on the Columbia, BC Electric backed off, stating its long-term interest in the Fraser but insisting that no development would occur before scientists could solve the fish-power problem. Throughout the 1950s, W. A. C. Bennett avoided a direct judgment on the Fraser River issue by consistently claiming that he would not support power development until the fish–power problem had a scientific solution.

By altering the emphasis of the fish–power debate and resisting the deployment of science toward a development agenda, BC fisheries scientists took a different approach than their American counterparts on the Columbia in the 1930s and after. Whereas in Canada fisheries scientists insisted on predevelopment studies and withheld approval of numerous dam projects, U.S. fisheries scientists, under different political pressures and without the benefit of hindsight, had helped to legitimize dam development by taking active roles in dam construction and planning and in boosting their own capacity to save runs by means of fishways and artificial propagation. This role was in part related to the long-standing dependence of fisheries science in the United States, particularly within the federal government, on funding tied to hatchery work. It was in light of this experience that BC fisheries scientists and American scientists within the IPSFC sought to insist that these earlier so-called solutions were no solutions at all and that understanding of the relevant problems only became more complex over time.

Apart from shifting the discourse of the fish vs. power debate and positioning science within it, fisheries scientists found it possible to play a more active political role in resisting specific development proposals. As discussed in Chapters 5 and 6, scientists gained powerful positions within the Department of Fisheries, the IPSFC, and the Fraser Basin Board to assess dam projects and comment on their merits and demerits during the planning stage. Scientists played more than the role of advisors in this capacity because of the legal authority of the Department of Fisheries under Section 20 of the Fisheries Act, which allowed the department to insist on remedial measures in river structures. Although, as in the Alcan case, scientific assessments could be ignored or overridden, even in this case, some important and expensive changes were made to the original project design. Scientists could not simply deny development projects, but they could suggest imposing expensive mitigation exercises on developers or place potential projects in such a poor light as to produce political difficulties.

The influence of science on the fish vs. power debate was not unidirectional. Politics changed science and its knowledge. Throughout the 1950s, the search for a solution to fish and power – both in fundamental and applied studies – occurred within a new institutional framework. In response to this looming problem, funding for fisheries science increased, institutions emerged, and cooperative linkages were created. These shifts had variable impacts. UBC probably benefited the most from expanded funding; the Department of Fisheries became the institution with the most porous boundaries and strongest cooperative links and coordinating function. Scientists in these institutions made and gained new roles within the Pacific fisheries science community as a result.

The outcomes of research had more significance and meaning than the instrumental goals to which some wished to direct them. In 1970, Peter Larkin reflected on the course of research during the 1950s and observed, "From a biological point of view, the fish–power problem spurred interest in physiology and behaviour, bringing to light a better appreciation of the many adaptations of salmon to their environment and mode of life."[98] In 1956, William Hoar stated that the fish–power debate had acted as a forcing ground for drawing fish behaviorists into the broader community of fisheries science; political pressures erased former disciplinary barriers.[99] The cooperative project of fish–power research and the intersecting of different intellectual avenues of approach produced significant side effects for the development of biological thought. What remains unstated in Larkin's observation, of course, is that the focus on problems of inland waters and the questions of changing freshwater environments produced a different and less salutary by-product in a concomitant inattention to problems of ocean migration and oceanography – areas of research that with new tools and techniques would become very important in subsequent decades. Although research surrounding the negotiations of the International Convention for High Seas Fisheries of the North Pacific Ocean signed with Japan and the United States in 1952 produced new knowledge of ocean habitats of salmon, the fish–power problem emphasized a bias toward riverine environments.[100] There is no doubt that the increased funding of applied problems on fish and power produced important fundamental findings about salmon life history as well; but it was also the case that the emphasis

98. Larkin, "Management of Pacific Salmon," p. 232.
99. UBC, Fraser River Hydro and Fisheries Research Project Papers, File 4, William S. Hoar, "Power Development and Fish Conservation on the Fraser," Technical Report – Biological, September 15, 1956.
100. On the impact of this convention, see: Healey, "The Management of Pacific Salmon Fisheries," p. 250.

of funding, institutional research programs, and scientists' political contexts directed science in ways that did not necessarily overlap with questions of a fundamental character. It may well be impossible to know, as Robert E. Kohler suggests in a study of scientific institutions, what would have happened to science without the impetus of institutional funding and the pressures of the political context.[101] But it is tempting to wonder: Given the freedom to pursue their own questions, would fisheries biologists have focused on the study of salmon responses to environmental change under dam development? Or would they have taken a broader view, encompassing the ocean and the river, and not looking to immediate ends, but to long-term topics of ecological significance?

101. Kohler, *Partners in Science*, pp. 2–3.

Conclusion

The free flow of the Fraser River bears the consequences of history but reveals none of its causes. Over the twentieth century, this river has played host to dreams of liberation and transformation, to physical changes and social consequences, to protective actions and inactions. Yet, against the predictions of most observers in the early and middle parts of the century, the river runs freely in its main course. The river plays host to dreams, but not to large dams.

From a comparative perspective, this outcome is surprising. In the regional context of western North America, the Columbia River, the Fraser's closest parallel case, bears the weight of sixteen main-stem dams. Among Canada's largest rivers, only the Mackenzie River remains, like the Fraser, undammed. Within BC, smaller rivers such as the Stikine, Nass, and Skeena have not been dammed on their main stems, but they contain much smaller power potential, lie at a distance from major centers of population, and provide habitat, like the Fraser, for major salmon runs. Over the twentieth century, Canadians have dammed rivers across northern North America from the Saguenay to the St. Lawrence to the Saskatchewan, and executed over fifty interbasin transfers, some on a massive scale.[1] In the northern third of the world, according to Dynesius and Nilsson, there are only five other rivers of comparable size with the Fraser that experience little or no fragmentation in their main channels.[2] In North America, the Yukon River provides

1. Woo, "Water in Canada, Water for Canada," p. 87.
2. Dynesius and Nilsson, "Fragmentation and Flow." These rivers all have a virgin mean annual discharge (VMAD) above $3200 m^3 s^{-1}$. Dynesius and Nilsson define VMAD as "the discharge before any significant direct human manipulations"(p. 753).

a very rough parallel, although in an utterly different social, economic, and environmental context. In Eurasia, the Khatanga, Pechora, Amur, and Lena Rivers remain undammed, but, unlike the Fraser, they lie at a considerable distance from densely settled regions. Some flow through the Arctic and freeze for part of the year. Viewed comparatively, the Fraser cannot help but appear unique. There are few other rivers around the world as large and capable of dam development as the Fraser that remain, nevertheless, undammed. Why is this so?

Driving Development

The answer is not that development pressures were not felt on the Fraser River. In some respects, development on the Fraser simply replicated international patterns. But the nature of that replication, its local effects, did matter. Early dam development imported approaches and patterns, particularly from the British Empire and the United States. A private electrical company backed by British investors with an expertise in international utilities applied new hydroelectric technologies to lower-basin tributaries. Resource firms in the Interior raised dams that would not have been out of place – in their design or purpose – anywhere on the West Coast of North America. Federal and provincial scientists measured rivers, analyzed their characteristics, and spoke the idiom of what Donald Worster calls the "global engineering priesthood."[3] In these respects, the Fraser River and BC looked like many places around the globe that sought to adopt British and American advances in dam technology and apply them locally.

If BC sought to capture some of these international advances and employ them as the basis for regional economic growth, then it did so only haltingly. During the 1930s, after the basic pattern of hydroelectric development in the Fraser Basin had been established, contracting commercial and consumer electrical demand led the province's major utility to suspend expansion and consolidate the existing system. To the south, by contrast, on the American

3. Worster, *An Unsettled Country*, p. 36.

Columbia main-stem development began. A federal government bent on demand stimulation, navigation improvements, irrigation development, and public works pressed forward New Deal projects to transform the river into a regional power generator. In Canada, no New Deal program created the possibility for state-led development. In BC, navigation beyond the lower river was not feasible, irrigation lobbies were relatively marginal, and power development remained firmly in the hands of private interests with little appetite for economically risky dam projects.

During World War II, the comparative consequences of this development pattern were not lost on British Columbians. As industrial development expanded in the U.S. Pacific Northwest, British Columbians faced restricted power supplies and occasional brownouts in urban centers. Public demands for new power sources gained strength. Criticism of delayed development took hold. BC entered a game of catch-up familiar to peripheral economies around the world. By the late 1940s, the provincial government had applied sufficient pressure to BC Electric that the private utility pursued a development agenda on the Bridge River, a middle-basin tributary. A public agency, modeled on central Canadian and American institutions of rural and public electrification, extended electricity to hinterland regions. A major flood in 1948 suggested the need for integrated river planning, and a joint federal–provincial board pursued measures to avoid future flood damage, taking inspiration in spirit, if not in institutional power, from the TVA. By the late 1940s, the Fraser remained less developed than the Columbia, but, increasingly, Canadians embraced the American development model as the path to future prosperity.

On a global scale, the 1950s was a critical decade of dam development. In a period of marked economic growth, dams became a primary building block for agricultural expansion into semiarid and arid lands and as a generator for regional industrialization. The scale of development, however, also owed something to the intensifying conflicts of the Cold War. Dams suited diverse ideological agendas to model political power and combat perceived distant threats. Superpowers funded dam development abroad to inspire allegiance. In BC newspapers, editorialists asked why this place did not produce dams as big as those of the Reds. It was a sentiment that might have been offered in numerous western bloc states.

BC entered this period with a number of characteristics recommending it well for dam development. It offered industrial developers cheap waterpower, access to North American markets, and political stability. Alcan took advantage of the opportunity, raised a massive upper-basin diversion scheme, and developed a large settlement and smelter facility on the coast. It pursued a production model already developed in eastern North America on the Saguenay River and in western North America on the Columbia. The firm bet successfully on the prospect of increased North American aluminum demand in the context of Korean War and Cold War military spending. It also demonstrated the depth of Cold War anxieties by burying the project's powerhouse within the granite housing of Mount Dubose, in part to protect against Soviet bombing.

By the middle 1950s, transnational forces increasingly shaped the development agenda. Local elites that sought to gather political support for a series of main-stem dams on the Fraser identified American interests as the major financial backers. Attempts to broker an agreement with the United States to dam the upper Columbia implicated the Fraser. General McNaughton, Canada's IJC chair, sought to divorce Canada's water-development future from the United States by proposing to divert the Columbia into the Fraser. His suggestion helped to put pressure on the United States to meet Canadian negotiating demands, but it gained only limited domestic support and active provincial criticism. At the same time, Swedish investment capital opened the question of northern development on the Peace River. By the early 1960s, Canada and the United States agreed to integrate development on the Columbia. As a result, BC pressed forward to develop both the Peace and Columbia and market power under a newly created provincial corporation, BC Hydro. On the Fraser, no coordinated power lobby emerged, only a series of competing interests. As complicated as this development pattern was by local, regional, national, and international pressures, the development decade in BC was but one aspect of a broader international trend. The critical variation was this: The Fraser remained undammed.

Constraining Development

If BC followed so many basic patterns of twentieth-century international water development, then what accounted for this unusual outcome on the Fraser? What, in short, constrained development on the Fraser?

Before 1945, there were few constraints on dam development on the Fraser River. The authority of the federal Department of Fisheries to limit and modify dam development on salmon streams was rarely exercised, and when it was, it lacked technical competence. In the lower basin, dam development remained tied to tributaries, not because it was blocked on the river's mainstem, but because local market demand would not support a large dam project, and suitable dam sites were too distant to provide efficient and affordable transmission. Although one might have expected the river's most famous obstruction at Hells Gate to provide a warning against the dangers of main-stem dam development, its effects were not well understood before World War II. Hells Gate shaped the fisheries, but it had not yet become an emblem of the potential consequences of dams.

By World War II, however, just as British Columbians began to imagine a new day of power, an international commission of fisheries scientists began to restore the river's salmon. The Pacific Salmon Convention of 1937 established a commission to conduct a basinwide survey of Fraser sockeye salmon. Commission scientists focused on Hells Gate as a continuing obstruction to salmon migration and sought to reconstruct the site with fishways. In so doing, they also highlighted the difficulties posed to salmon by main-stem obstructions. Hells Gate was compared to Columbia River dams. British Columbians were reminded that things could be worse. The actions of the commission not only restored the salmon, but also laid a claim for fisheries interests to the river as salmon spawning habitat.

The fisheries claim was challenged in the immediate postwar period when Alcan proposed a major dam project, either on Chilko Lake or the Nechako River. Unlike before the war, the federal Department of Fisheries took an active role. Its officials went further and organized a unified fisheries defense. Fisheries scientists modeled consequences. After some hesitation, fishing industry groups collaborated to protest development, at least on Chilko Lake, the more productive salmon river of the two. Faced with a major industrial interest, backed by a wide public and unstinting provincial support, leading canners brokered a deal to block development on Chilko Lake in exchange for acceptance of development on the Nechako. The constraint, in this instance, was only partial. When fisheries scientists and federal officials attempted to propose modifications to the Alcan project to mitigate effects on salmon migration, they found little success.

The Alcan case acted as a catalyst for the formation of a fisheries protection coalition. Future projects that proposed main-stem dams at Moran, Columbia-to-Fraser diversions, and other such schemes were met by a reasonably coherent voice of protest. This coalition depended on the organizing capacities of the federal Department of Fisheries and the active involvement of otherwise divergent groups: capitalist canners, communist unionists, members of fisheries cooperatives, native fishers, sport fishers, and fisheries scientists. In local disputes, in the press and the public sphere, this coalition railed against the easy association of power and progress. Fisheries defenders sought to invoke a rhetoric of salmon and place and of salmon and heritage. They also positioned salmon as a critical element in the Cold War. Protecting fisheries, they argued, could go hand in hand with future atomic energy development; fisheries supplied a food reserve in the event of nuclear warfare. It is also true that fisheries interests could point to the American experience on the Columbia to defend their dire warnings. The rapid declines of salmon runs on the Columbia after the 1930s hung like a shadow over the Fraser, as one journalist said. The earlier American experience informed and reinforced the Canadian fisheries defense. From a narrow industry base, the fisheries coalition gained wide support and helped to hold development off the river.

The formation of a coalition of this kind was both locally important and internationally anomalous. On the international scene, antidam protest movements were limited to local and national contexts before the 1970s.[4] In these settings, and particularly in the United States, they focused primarily on wilderness preservation,

4. McCully, *Silenced Rivers.*

except on the Columbia where commercial fishers had protested dams since the 1920s.[5] In the Canadian case, however, the fisheries coalition contained American and Canadian interests; it also built from an industrial conservationist rather than a preservationist base of support. Compared with fisheries protest against dam development on the American Columbia, the Canadian fisheries defense was also remarkably coherent. In the U.S. Pacific Northwest, divisions based on ethnic background, gear type, and location in the context of a declining fishery diminished political cooperation. In Canada, for a time, these same divisions were transcended. In 1956 one American fisheries lobbyist observed, "I understand the fisheries people of British Columbia are making a much stronger fight against power dams on salmon streams than the Americans have been making down here in connection with the Columbia River which is gradually, year by year, being destroyed as the great salmon river."[6]

The comparative role of state agencies was also important. Although several fisheries and wildlife agencies such as the U.S. Fish and Wildlife Service protested development on the Columbia, they faced considerable opposition from bureaucratic opponents like the U.S. Army Corps of Engineers, the Bureau of Reclamation, and the BPA. Further, the emphasis placed on fish culture activities within the Western American fisheries science community made some of its members predisposed to assist development activities with fish cultural and mitigation experimentation.[7] On the Fraser, by contrast, fisheries agencies actively organized a fisheries defense to foment protest, reaching beyond bureaucratic politics to the broader civil society. No coherent prodevelopment bureaucratic bloc emerged to counter the role of the Department of Fisheries, despite supportive actions at both the provincial and the federal levels. Canadian fisheries scientists, with little experience in fish culture and a body of experimentation critical of its effects, refused to offer partial solutions to the problem of fish and dams and actively participated in the fisheries coalition.

5. Harvey, *A Symbol of Wilderness*.
6. UBC Special Collections and Archives, Doyle Henry, Papers, Box 1, File 1-2, E. D. Clark, secretary treasurer of the Association of Pacific Fisheries, Seattle, to Doyle, May 4, 1956.
7. Taylor, *Making Salmon*; idem, The Political Economy of Fishery Science"; White, *The Organic Machine*.

The ability of BC's fisheries coalition to block development on the Fraser depended critically on another factor: the presence of alternatives. It is difficult to imagine that dam development would not have proceeded on the Fraser in the 1950s or 1960s had there been no other large river development opportunities within the same political jurisdiction. Because of international interest in the upper Columbia and the Peace and advances in long-distance transmission technology, some of the barriers to development in the Interior and North were beginning to disappear. It was imaginable that electricity could be transmitted from the provincial North to the metropolitan South in 1958 in a way it had not been a decade before. Financing large-scale development appeared feasible by the early 1960s because of the willingness of the United States to pay downstream benefits to Canada in exchange for the construction of upper-basin storage. Given the political difficulties of development on the Fraser, the provincial government pursued a strategy to develop the Interior and North. Exporting the environmental consequences of large dam development to other parts of the province and beyond suited provincial politicians eager for modern dam development and increased electrical supplies, as well as for a political program that would not alienate fishing interests and their supporters.

The trajectory of this development pattern was reinforced in important ways by the influence of the United States. U.S. fishing interests held a strong stake in preserving the Fraser as salmon spawning habitat. Under the Pacific Salmon Convention, they reaped half of the river's sockeye salmon. The United States government, furthermore, had helped to fund restoration and conservation measures at Hells Gate and other sites on the Fraser River. Fraser development threatened American interests. In the Interior, the United States was also concerned about the future course of development on the Columbia. Since the mid-1940s, the United States had sought to develop an integrated development plan with Canada to regulate downstream flows and guard against the flood threat. Canadian intransigence to this agenda threatened American power-development potential and industrial interests. The prospect of a Columbia-to-Fraser diversion also could have damaged the Fraser fishery and therefore the American fishing industry. By fighting development on the Fraser, the American government helped to limit Canadian options on the Columbia. By pressing for coordinated development on the Columbia, it also diminished the potential or rationale for development on the Fraser. Although Canadians are wont to compare the case of the Canadian Fraser and that of the American Columbia and congratulate themselves for environmental foresight, it would be more accurate to conclude that the Fraser was not dammed in spite of Canadian attempts by a conjunction of Canadian and American interests.

Retrospect and Prospect

Whether or not the conditions that held dams off the Fraser River for most of the twentieth century will remain in the future is difficult to determine. Since the late 1960s, several attempts to dam the Fraser have appeared. Moran became the focus of renewed interest in the early 1970s. Diversions from the upper Fraser to the Peace basin have also been investigated. In the late 1980s and 1990s, Alcan attempted to expand its Nechako diversion project, but met with considerable public resistance; the provincial government intervened and canceled the plan.

In the 1950s and 1960s, a strong fisheries coalition, alternative development sites, and American influence helped to keep dams off the Fraser. Today, the fishing industry has contracted, along with salmon supplies. Former Canadian and American cooperation in the regulation of the Pacific salmon fishery has faltered; salmon wars mark the press. No other large, undeveloped rivers, free from the politics of salmon, exist in the province to offset threats on the Fraser. The United States now plays a significant role as an importer of Canadian electricity. The thin edge of the Columbia River Treaty has developed into a wedge of the Canadian export economy. Could expanding American energy demands, in an increasingly continental trade regime, force future development on the Fraser? Perhaps.

In contrast to the fish vs. power era, however, new constraints on development have appeared. Since the 1970s, the politics of native land claims and resource access, particularly in the fisheries, have resurfaced with force. Whereas native peoples were barely consulted about the effects of development during the fish vs. power era, today it is inconceivable that attempts to dam the Fraser would not be met by a coherent and strong native defense, backed by significant legal challenges and considerable public support. In addition, whereas the environmental movement was only beginning to take form in the early 1970s in BC, it is now a potent political force. Its influence, admittedly, has not been felt at the ballot box, but it has shaped the direction of public policy and, more broadly, political culture. In 1997, the provincial government passed legislation that explicitly prohibits main-stem development on the Fraser.[8] If the fish vs. power debate were to reemerge in the coming decades, organized opposition groups and existing legislation would make the development agenda difficult to achieve.

The case of the Fraser offers some hope about the possibility of balancing development and environment in the future. Against broad international trends, Canadians have taken a different path on the Fraser. They have done so, however, without choosing between fish and power, but by choosing fish and power.

8. British Columbia, Bill 25 (1997), *Fish Protection Act.*

Bibliography

I. Primary Sources

1. Archival Documents

i. British Columbia Archives and Records Service (BCARS)
Ad MSS 775, Henry Forbes Angus, *My First Seventy-Five Years*
Ad MSS 1147, S. H. Frame Papers
Ad MSS 392, Frank Swannell Papers
Ad MSS 1977, Survey Diary, Chilko Lake, 1929
Ad MSS 2625, Taylor, W. Diary
Ad MSS 2812, E. D. Taylor Papers
BCARS, Photograph Collection
GR 1378, BC Commercial Fisheries Branch
GR 442, BC Energy Board
GR 1390, BC Energy Commission
GR 1027, BC Fish and Game Branch
GR 435, BC Department of Fisheries
GR 1027, BC Fish and Wildlife Branch
GR 1118, BC Marine Resources Branch
GR 880, BC Power and Special Projects
GR 1222, BC Premiers' Papers
GR 1414, BC Premiers' Papers
GR 1160, BC Public Utilities Commission
GR 1427, BC Water Management Branch
GR 884, BC Water Rights Branch
GR 1006, BC Water Rights Branch
GR 1236, BC Water Rights Branch
GR 1289, BC Water Rights Branch
GR 123 Canada, Department of Indian Affairs, BC Records

ii. City of Vancouver Archives
Ad MSS 256, Archibald, Harry P. Papers

Ad MSS 321, BC Electric Co. Ltd.
Ad MSS 530, Smith, Arthur G. Papers
Ad MSS 69, Stevens, H. H. Papers

iii. Crown Lands Registry, Victoria, BC
Department of Lands 'O' Files

iv. National Archives of Canada (NAC), Pacific Region
RG 23 Department of Fisheries

v. National Archives of Canada (NAC)
RG 15, Department of the Interior
RG 23, Department of Fisheries
RG 51, International Joint Commission
RG 89, Water Resources Branch

vi. Pacific Salmon Commission Library and Archives (PSCA)
Unpublished Reports and Correspondence Files

vii. Simon Fraser University (SFU) Archives
W. A. C. Bennett Papers

viii. University of British Columbia Special Collections and Archives (UBC)
BC Electric Railway Papers
Doyle, Henry Papers
Fisheries Association of BC Papers
Fraser, Charles MacLean Papers
Fraser River Hydro and Fisheries Project Papers
Haig-Brown, Roderick Papers
IPSFC Papers
Larkin, Peter A. Papers
Mackenzie, Norman A. M. Papers
Pretious, Edward S. Papers
Scott, Anthony Papers
Sinclair, James Papers
Tolmie, Simon Fraser Papers
United Fishermen and Allied Workers Union Papers
Warren Family Papers

ix. University of Victoria Special Collections and Archives
Mayhew, Robert Papers

x. University of Washington Archives (UWA)
Acc. 129-3, Allen, E. W. Papers
Acc. 861-1, Doyle, Henry Papers
Acc. 1038, Freeman, Miller Papers
Acc. 2597-77-1, 2597-3-83-21, Thompson, W. F. Papers
Acc. 1683-71-10, 1683-4-85-4, Van Cleve, Richard Papers

xi. Water Management Branch, Victoria, BC
Department of Lands and Forests 'O' Series Files

2. Newspapers and Annual Reports

A note on newspaper articles and annual reports cited: For the most part, this study relies on newspaper sources collected by institutions in clippings files. I have done some original searches in the press to verify stories and dates that seemed particularly pertinent to my research; however, without the existence of well-documented clipping files maintained by Canners' groups for the early part of the twentieth century and BC Electric for the mid-century, my research in the press could not have been as broadly based. Both the canners' scrapbooks and the BCER clippings file are housed at the UBC Special Collections and Archives. In the body of the study, I have cited the relevant collection before newspaper references in shortened form (CSB and BCER CF). I have also read the annual reports of a host of different government departments, institutions, and organizations for the years covered in this study. In the interests of brevity and clarity, I have refrained from including all of these in the bibliography and list only those cited in the main text.

3. Printed Documents

No author, but probably Ashdown Green. "The Salmonidae of British Columbia." *Papers and Communications Read Before the Natural History Society of British Columbia* 1(1) Victoria: Jas. A Cohen Printer, 1891.
   "The Decline of the Sockeye." *Industrial Progress and Commercial Record* V(4) (1917): 385–90.
"Do Nothing Biology." *Pacific Fisherman*, 45(7) (1947): 30.
"Forum: Fish and Power." *Transactions of the Fourth British Columbia Natural Resources Conference* (BC Natural Resources Conference, 1951): 95– 150.
"Hell's Gate and the Sockeye." *Commercial Fishermen's Weekly* XIII (8) (March 14, 1947): 90–1.
House of Commons. *Debates.*
Industrial Development Act, 1949, *Statutes of British Columbia*, Chapter 31.
Interim Report of the Post-War Rehabilitation Council. Hon. H. G. T. Perry, Chair, Victoria, January, 1943.
Joint Hearings Before the Committee on Interior and Insular Affairs and a Special Subcommittee of the Committee on Foreign Relations United States Senate Eighty-Fourth Congress Second Session, March 22, 26, 28, and May 23, 1956. Washington, D.C.: Government Printing Office, 1956.
Nature's Fury: The Inside Story of the Disastrous BC Floods (May-June 1948) held at BCARS.
"Preparing to Open Hell's Gate." *Pacific Fisherman* 43(1) (1945): 63.
"Research Board Said Not Open to Charges." *Commercial Fishermen's Weekly* XIII (13) (April 25, 1947): 152–3.

"Review of Evidence Suggested by Ricker." *Commercial Fishermen's Weekly* XIII (12) (April 18, 1947): 135–7.

"Salmon Blockade." *Vancouver Sun*, September 21, 1941.

"Salmon Board Declares Hells Gate Must Be Cleared." *Vancouver Sun*, December 5, 1941.

"Salmon Commission Hits Back at Critic." *Commercial Fishermen's Weekly* XIII (10) (March 28, 1947): 111, 113.

"Scientific Sharpshooting." *Pacific Fisherman* 45 (5) (1947): 37.

The British North America Act, 1867, *Statutes of the United Kingdom*, 30–31, Victoria, Chapter 3.

"The Electric Power Act Outlined in an Address given by the Honourable John Hart Premier of BC in Moving Second Reading of Bill, Thursday, March 15, 1945, during Session of the Legislature."

Aluminum Company of Canada. "Kitimat-Kemano: Five Years of Operation, 1954–1959." Vancouver: Aluminum Company of Canada, Limited, 1959.

Andrew, F. J. and G. H. Geen. *Sockeye and Pink Salmon Production in Relation to Proposed Dams in the Fraser River System.* Bulletin XI. New Westminster: IPSFC, 1960.

Babcock, John Pease. *Report of the Fisheries Commissioner for British Columbia for the year 1902.* (1903): 1–35.

"The Spawning Beds of the Fraser." In *Report of the Fisheries Commissioner for British Columbia for the year 1913* (1914): 20–38.

Bell-Irving, H. B. "Conditions in the Fraser Canyon – A Canadian View." *Pacific Fisherman* 28(8) (1930): 16–17.

Bell, Milo C. "Report on the Engineering Investigation of Hell's Gate, Fraser River." *Annual Report of the IPSFC* (1944): 15–22.

"Fisheries Research at High Dams in Washington State." *Transactions of the Sixth British Columbia Natural Resources Conference* (Victoria: BC Natural Resources Conference, 1953): 102–4.

Brett, J. H. "The Nature of the Biological Problem." *Transactions of the Sixth British Columbia Natural Resources Conference* (Victoria: BC Natural Resources Conference, 1953): 96–102.

British Columbia. *Report of the Commissioner of Fisheries for British Columbia for the year 1913* (1914): 1–148.

Progress Reports of the Rural Electrification Committee. Victoria: King's Printer, 1944.

Report of the Commissioner of Fisheries for British Columbia for the year 1917 (1918): 1–126.

Report of the Commissioner of Fisheries for British Columbia for the year 1921 (1922): 1–79.

Carpenter, E. E. "The Water Developments of the Alouette–Stave–Ruskin Group of the British Columbia Electric Railway Company, Limited." *Engineering Journal* X(1) (1927): 17–18.

Clay, C. H. *Design of Fishways and Other Fish Facilities.* Ottawa: Department of Fisheries Canada, 1961.

"Problems Associated with Upstream Migration Over High Dams." *Transactions of the Sixth British Columbia Natural Resources Conference* (Victoria: BC Natural Resources Conference, 1953): 91–3.

Clay, C. H. and Peter Larkin, "Artificial Propagation is NOT the Answer to the Problem of Fish and Power on the Fraser." *BC Professional Engineer* 10(3) (1959): 20–3.

Clemens, W. A. "Some Historical Aspects of the Fisheries Resources of British Columbia." *Transactions of the Ninth British Columbia Natural Resources Conference* (Victoria: BC Natural Resources Conference, 1956): 119–30.

Crippen Wright Engineering Ltd. *Electric Power Requirements in the Province of BC.* Vancouver, April, 1958.

Davis, Arthur V. *Water Powers of Canada.* Ottawa: Commission of Conservation, 1919.

Davis, E. and E. G. Marriott. "British Columbia Dams." *Engineering Institute of Canada Transactions* 35 (1923–5): 135–41.

Department of Trade and Commerce. *Electric Power Demand and Supply, British Columbia, 1929 to 1980.* Ottawa: 1957.

Dominion of Canada. *The Fisheries Acts (1868).* Ottawa: Department of Marine and Fisheries, 1873.

Farrow, R. C. "Snow Surveys for the Purpose of Forecasting Streamflow." *Forestry Chronicle* Vol XIII (1) (1937): 1–15.

"Snow Surveys: A New Medium for Forecasting Run-Off." *Engineering Institute of Canada Transactions* XXI(10) (1938): 451–5.

"Forecasting Run-Off from Snow Surveys." *The Geographical Journal* C(5–6) (1942): 204–18.

"The Search for Power in the British Columbia Coast Range." *Geographical Journal* CVI(3–4) (1945): 89–117.

Foerster, R. E. "A Comparison of the Natural and Artificial Propagation of Salmon." *Transactions of the American Fisheries Society* 61 (1931): 121–30.

"Comparative Studies of the Natural and Artificial Propagation of Sockeye Salmon." *Proceedings of the Fifth Pacific Science Congress Canada, 1933* V (1934): 3593–7.

Fraser River Board. *Preliminary Report on Flood Control and Hydro-Electric Power on the Fraser River.* Victoria, 1958.

Final Report on Flood Control and Hydro-Electric Power in the Fraser River. Victoria, 1963.

Freeman, Otis W. "Salmon Industry on the Pacific Coast." *Economic Geography* II(2) (1935): 109–29.

Geen, G. H. and F. J. Andrew, "Limnological Changes in Seton Lake Resulting from Hydroelectric Diversions." IPSFC Progress Report No. 8 (Vancouver: IPSFC, 1961).

Gordon, R. N. "Fisheries Problems Associated with Hydroelectric Development." *Engineering Journal* 47(10) (1964): 31–7.

Gwyther, Val. "Multiple Purpose Development of the Fraser River Basin." *BC Professional Engineer* 9(10) (1958): 13–9.

"Watershed Resource Value: The Associated Development of Fish and Power."
  Engineering Journal 44(11) (1961): 49–56.

Hamilton, J. A. R. and F. J. Andrew. *An Investigation of the Effect of Baker Dam
  on Downstream-Migrant Salmon.* Bulletin VI. New Westminster: IPSFC,
  1954.

Hutchison, Bruce. "International Sockeye Board Inspects Fraser River Blockade."
  Vancouver Sun, September 21, 1941.

IPSFC. *A Review of the Sockeye Salmon Problems Created by the Alcan Project
  in the Nechako River Watershed.* New Westminster: IPSFC, 1953.

  "Hell's Gate Fishways." (pamphlet) New Westminster, 1971.

Jackson, R. I. *Variations in Flow Patterns at Hell's Gate and Their Relationships
  to the Migration of Sockeye Salmon.* Bulletin III. New Westminster: IPSFC,
  1950.

  "Measurement of Losses of Fingerling Salmon at High Dams." *Transactions
  of the Sixth British Columbia Natural Resources Conference* (Victoria: BC
  Natural Resources Conference, 1953): 93–6.

Jomini, Harry. "The Kenney Dam." *The Engineering Journal* 37 (11) (1954):
  1386–97.

Jordan, David Starr. *A Guide to the Study of Fishes.* New York: Holt, 1905,
  Vol. II.

Kent, T. W. "Power for Aluminum." *The Beaver* 283 (1953): 4–9.

Larkin, Peter A. *Power Development and Fish Conservation on the Fraser River.*
  Vancouver: Institute of Fisheries, UBC, 1956.

  ed. *The Investigation of Fish-Power Problems.* Vancouver: Institute of Fisheries, UBC, 1958.

McHugh, J. H. "Report on the Work of Removal of Obstructions to the Ascent of
  Salmon on the Fraser River at Hell's Gate, Skuzzy [sic] Rapids, China Bar
  and White's Creek during the Year 1914, and the early portion of the year
  1915." In *Annual Report, Fisheries Branch, Department of Naval Service
  1914–15* (1915): 263–75.

Meurling, H. F. "Description of Work at Hydrographic Station Near Nelson."
  Annual Report Department of Lands 1912 (1913): 143–5.

Muir, J. F. and Eugen Ruus. "Engineering Research on the Fish Power Problem."
  Engineering Journal 44(10) (1961): 98–108.

Napier, G. P. "Report on the Obstructed Conditions of the Fraser River at Scuzzy
  Rapids, China Bar, Hell's Gate, and White's Creek." In *Report of the
  Commissioner of Fisheries for British Columbia for the year 1914* (1915):
  39–42.

Perry, H. G. T. *Interim Report of the Post-War Rehabilitation Council.* Victoria,
  January 1943.

Potter, R. E. "Moran Dam – Fish and Power." *BC Professional Engineer* 9(10)
  (1958): 21–8.

Ricker, William E. "Hell's Gate and the Sockeye." *Journal of Wildlife Management* 11(1) (1947): 10–20.

  "Effects of the Fishery and of Obstacles to Migration on the Abundance of

Fraser River Sockeye." *Canadian Technical Report of Fisheries and Aquatic Sciences* (1987), No. 1522.

Rounsefell, George A. and George B. Kelez. "The Salmon and Salmon Fisheries of Swiftsure Bank, Puget Sound, and the Fraser River." *Bulletin of the Bureau of Fisheries*. Washington D.C.: U.S. Department of Commerce, 1938, Bulletin No. 27, Vol. XLIX, pp. 693–823.

Royal, Loyd A. "The Rebirth of the Fraser Sockeye in Dollars and Sense." *Transactions of the Seventh British Columbia Natural Resources Conferences* (Victoria: BC Natural Resources Conference, 1954): 11–4.

Shaw, John B. "The BC Power Commission and the Development of Rural Electrification." *BC Professional Engineer* 1(3) (1950): 7–10.

Steede, J. H. "The Long Distance Transmission of Energy." *BC Professional Engineer* 3(5) (1953): 16–20.

Talbot, G. B. *A Biological Study of the Effectiveness of the Hell's Gate Fishways*. New Westminster: IPSFC, 1950, Bulletin III.

Thompson, William F. *Effect of the Obstruction at Hell's Gate on the Sockeye Salmon of the Fraser River*. New Westminster: IPSFC, 1945, Bulletin 1.

Walker, W. M. "The Cost of Electrical Energy Generation and Transmission." *BC Professional Engineer*, 5(12) (1954): 20–4.

Warren, Harry V. "National and International Implications Involved in the Development of a Portion of the Lower Fraser River." *Transactions of the Fifth British Columbia Natural Resources Conference* (Victoria: BC Natural Resources Conference, 1952): 257–68.

"The Moran Dam." *Canadian Mining Journal* 80(3) (1959): 63–8.

White, Arthur V. *Fishways in the Inland Waters of British Columbia*. Ottawa: Commission of Conservation, Canada, 1918.

Water Powers of British Columbia. Ottawa: Commission of Conservation, 1919.

II. Secondary Sources

Allen, Cain. "'They Called It Progress': Indians, Salmon, and the Industrialization of the Columbia River." MA thesis, Portland State University, Portland, OR, 2000.

Amaral, Juan Carlos Gomez. "The 1950 Kemano Aluminum Project: A Hindsight Assessment." MA thesis, Simon Fraser University, Burnaby, BC, 1986.

Andrews, G. Smedley. "Major Richard Charles Farrow, B.C.L.S., P. ENG., 1892–1950." *The Link* 13(1) (1989): 3–6.

Barham, Bradford. "Strategic Capacity Investments and the Alcoa–Alcan Monopoly, 1888–1945." In *States, Firms, and Raw Materials: The World Economy and Ecology of Aluminum*, ed. Bradford Barham, Stephen G. Bunker, and Denis O'Hearn. Madison: University of Wisconsin Press, 1994, pp. 69–110.

Beamish, Richard J. and D. R. Bouillon. "Pacific salmon production trends in relation to climate," *Canadian Journal of Fisheries and Aquatic Science* (50) (1993): 1002–16.

Black, Michael. "Tragic Remedies: A Century of Failed Fishery Policy on California's Sacramento River." *Pacific Historical Review* LXIV(1) (1995): 37–70.

Bocking, Stephen. "Fishing the Inland Seas: Great Lakes Research, Fisheries Management, and Environmental Policy in Ontario." *Environmental History* 2(1) (1997): 52–73.

Buhler, Katharine. "Come Hell and High Water: The Relocation of the Cheslatta First Nation." MA thesis, University of Northern British Columbia, Prince George, BC, 1998.

Bunker, Stephen G. and Paul S. Ciccantell, "The Evolution of the World Aluminum Industry." In *States, Firms, and Raw Materials: The World Economy and Ecology of Aluminum*, ed. Bradford Barham, Stephen G. Bunker, and Denis O'Hearn. Madison: University of Wisconsin Press, 1994, pp. 39–62.

Cameron, Eion M. "Hydrogeochemistry of the Fraser River, British Columbia: Seasonal Variation in Major and Minor Components." *Journal of Hydrology*, 182 (1996): 209–25.

Campbell, Peter. "'Not as a White Man, Not as a Sojourner': James A Teit and the Fight for Native Rights in British Columbia, 1884–1922." *Left History* 2(2) (1994): 37–57.

Carlson, Roy L. "The Later Prehistory of British Columbia." In *Early Human Occupation in British Columbia*, ed. Roy L. Carlson and Luke Dalla Bona. Vancouver: University of British Columbia Press, 1996, pp. 215–26.

Cole, Douglas. "Leisure, Taste and Tradition in British Columbia." In *The Pacific Province: A History of British Columbia*, ed. Hugh J. M. Johnston. Vancouver: Douglas & McIntyre, 1996.

Church, Michael. "The Future of the Fraser River: Thinking About the River in the Lower Mainland." *Discover Magazine* (Fall 2002), www.naturalhistorybc.ca/VNHS/Discovery.

Copes, Parzival. "The Evolution of Marine Fisheries Policy in Canada." *Journal of Business Administration* 11(1/2) (1979/1980): 125–48.

Cronon, William. "Modes of Prophecy and Production: Placing Nature in History." *Journal of American History* 76(4) (1990): 1122–31.

Dorcey, Anthony H. J. "Water in the Sustainable Development of the Fraser River Basin." In *Water in Sustainable Development: Exploring Our Common Future in the Fraser River Basin*, ed. Anthony H. J. Dorcey and Julian R. Griggs. Vancouver: Westwater Research Centre, University of British Columbia, 1991, Vol. 2, pp. 3–18.

Drache, Daniel ed. "Celebrating Innis: The Man, The Legacy and Our Future." In *Staples, Markets and Cultural Change*, ed. Daniel Drache. Kingston and Montreal: McGill-Queen's University Press, 1995, pp. xiii–lix.

Dynesius, Mats and Christer Nilsson. "Fragmentation and Flow: Regulation of River Systems in the Northern Third of the World." *Science* 266 (1994): 753–61.

Ellis, Derek. "Construction – Hell's Gate (Canada)." In *Environments at Risk: Case Histories of Impact Assessment*. Berlin: Springer-Verlag, 1989, pp. 17–36.

Eng, Paula Louise. "Parks for the People? Strathcona Park 1905–1933." MA thesis, University of Victoria, Victoria, BC, 1996.

Evenden, Matthew. "Harold Innis, the Arctic Survey and the Politics of Social Science During the Second World War." *Canadian Historical Review* 79 (1) (1998): 39–68.

Forkey, Neil. "Maintaining a Great Lakes Fishery: The State, Science, and the Case of Ontario's Bay of Quinte, 1870–1920." *Ontario History* 87(1) (1995): 45–64.

Galois, R. M. "The Indian Rights Association, Native Protest Activity and the 'Land Question' in British Columbia, 1903–1916." *Native Studies Review* 8(2) (1992): 1–34.

Geen, Glen H. "Ecological Consequences of the Proposed Moran Dam on the Fraser River." *Journal of the Fisheries Research Board of Canada* 32(1) (1975): 126–35.

Gilhousen, P. "Estimation of Fraser River Sockeye Escapements from Commercial Harvest Data, 1892–1944." Vancouver: Pacific Salmon Commission, 1992.

Gladstone, Percy. "Native Indians and the Fishing Industry of British Columbia." *Canadian Journal of Economics and Political Science* XIX(1) (1953): 20–34.

Green, George. "Some Pioneers of Light and Power." *British Columbia Historical Quarterly* II(3) (1938): 145–62.

Haig-Brown, Roderick. "The Fraser Watershed and the Moran Proposal." *Nature Canada* 1(2) (1972): 2–10.

Hardwick, Walter. "The Effect of the Moran Dam on Agriculture within the Middle Fraser Region, British Columbia." MA thesis, University of British Columbia, Vancouver, BC, 1954.

Harris, Cole "Industry and the Good Life Around Idaho Peak." *Canadian Historical Review* 66(3) (September 1985): 315–43.

Hayden, Brian and June M. Ryder. "Prehistoric Cultural Collapse in the Lillooet Area." *American Antiquity* 56(1) (1991): 50–65.

Healey, M. C. "The Management of Pacific Salmon Fisheries in British Columbia." In *Perspectives on Canadian Marine Fisheries Management*, ed. L. S. Parsons and W. H. Lear. Canadian Bulletin of Fisheries and Aquatic Science 226 (1993): 243–66.

Hubbard, Jennifer. "Home Sweet Home? A. G. Hunstman and the Homing Behaviour of Canadian Atlantic Salmon." *Acadiensis* XIX(2) (Spring, 1990): 40–71.

Hudson, Douglas R. "Internal Colonialism and Industrial Capitalism." In *SA TS'E: Historical Perspectives on Northern British Columbia*, ed. Thomas Thorner. Prince George, BC: College of New Caledonia, 1989, pp. 177–213.

Keeling, Arn. "Ecological Ideas in the British Columbia Conservation Movement, 1945–1970." MA thesis, University of British Columbia, Vancouver, BC, 1998.

Keeling, Arn and Robert McDonald. "The Profligate Province: Roderick Haig-Brown and the Modernizing of British Columbia." *Journal of Canadian Studies* 36(3) (2001): 7–23.

Kennedy, Dorothy I. D. and Randy Bouchard. "Stl'atl'imx (Fraser River Lillooet) Fishing." In *A Complex Culture of the British Columbia Plateau: Traditional Stl'atl'imx Resource Use*, ed. Bryan Hayden. Vancouver: UBC Press, 1992, pp. 266–354.

Kew, Michael. "Salmon Availability, Technology, and Cultural Adaptation in the Fraser River Watershed." In *A Complex Culture of the British Columbia Plateau: Traditional Stl'atl'imx Resource Use*, ed. Bryan Hayden. Vancouver: UBC Press, 1992, pp. 177–221.

Kew, Michael and Julian R. Griggs. "Native Indians of the Fraser River Basin: Towards a Model of Sustainable Resource Use." In *Perspectives on Sustainable Development in Water Management: Towards Agreement in the Fraser River Basin*, ed. Anthony H. J. Dorcey. Vancouver: Westwater Research Centre, 1991, Vol. 1. pp. 17–47.

Klingle, Matthew. "Plying the Atomic Waters: Lauren Donaldson and the 'Fern Lake Concept' of Fisheries Management." *Journal of the History of Biology* 31 (1998): 1–32.

Kulik, Gary. "Dams, Fish, and Farmers: Defense of Public Rights in Eighteenth-Century Rhode Island." In *The Countryside in the Age of Capitalist Transformation: Essays in the Social History of Rural America*, ed. Steven Hahn and Jonathan Prude. Chapel Hill: University of North Carolina Press, 1985, pp. 25–50.

Larkin, P. A. "Management of Pacific Salmon of North America." In *A Century of Fisheries in North America*, ed. Norman G. Benson. Washington, D.C.: American Fisheries Society, 1970, Special Publication No. 7, pp. 223–236.

Lindsey, C. C. "Possible Effects of Water Diversions on Fish Distribution in British Columbia." *Journal of Fisheries Research Board of Canada* 14(4) (1957): 651–68.

Litvak, Isaiah A. and Christopher J. Maule. "Alcan Aluminum Ltd.: A Case Study" In *Royal Commission on Corporate Concentration*. Ottawa: Government of Canada, 1977, pp. 27–59.

Lovell, N. C., B. S. Chisholm, D. E. Nelson, and H. P. Schwarcz. "Prehistoric Salmon Consumption in Interior British Columbia." *Canadian Journal of Archaeology* 10 (1986): 99–106.

L'vovich, Mark I. and Gilbert F. White. "Use and Transformation of Terrestrial Water Systems." In *The Earth as Transformed By Human Action: Global and Regional Changes in the Biosphere Over the Past 300 Years*, ed. B. L. Turner II, William C. Clark, Robert W. Kates, John F. Richards, Jessica T. Mathews, William B. Meyer. Cambridge, England: Cambridge University Press, 1990, pp. 235–52.

MacCrimmon, Hugh. "The Beginnings of Fish Culture in Canada." *Canadian Geographical Journal* LXXI (3) (1965): 96–103.

Mathews, W. H. "From Glaciers to the Present." In *The Fraser's History*. Burnaby, BC: Burnaby Historical Society, 1977, pp. 9–18.

McFarlane, G. A., Richard S. Wydoski, and Eric D. Prince. "Historical Review of the Development of External Tags and Marks." *Fish Marking Techniques*, American Fisheries Society Symposium 7 (1990): 9–29.

McPhail, J. D. "The Origin and Speciation of *Oncorhynchus* Revisited." In *Pacific Salmon and Their Ecosystems: Status and Future Options*, ed. Deanna J. Stouder, Peter A. Bisson and Robert J. Naiman. New York: Chapman & Hall, 1997, pp. 29–38.

McVey, J. A. and J. E. Windsor. "The Value of Public Hearings as a Vehicle for Public Participation: A Case Study of the British Columbia Utilities Commission Review of the Kemano Completion Project." *Salzburger Geographische Arbeiten* 32 (1998): 81–120.

Mergen, Bernard. "Seeking Snow: James E. Church and the Beginnings of Snow Science." *Nevada Historical Society Quarterly* 35(2) (1992): 75–104.

Millerd, Frank. "Windjammers and Eighteen Wheelers: The Impact of Changes in Transportation Technology on the Development of British Columbia's Fishing Industry." *BC Studies* 78 (1988): 28–52.

Moore, R. Daniel. "Hydrology and Water Supply in the Fraser River Basin." In *Water in Sustainable Development: Exploring Our Common Future in the Fraser River Basin*, ed. Anthony H. J. Dorcey and Julian R. Griggs. Vancouver: Westwater Research Centre, UBC, 1991, Vol. 2, pp. 21–39.

Muckleston, Keith W. "Salmon vs. Hydropower: Striking a Balance in the Pacific Northwest." *Environment* 32(1) (1990): 10–15, 32–6.

Naske, Claus M. "The Taiya Project." *BC Studies* 91–92 (1991–2): 5–50.

New, W. H. "The Great River Theory: Reading MacLennan and Mulgan." *Essays in Canadian Writing* 56 (1995): 162–82.

Newell, Dianne. "The Politics of Food in World War II: Great Britain's Grip on Canada's Pacific Fishery." *Historical Papers 1987 Communications Historique*: 178–97.

"Dispersal and Concentration: The Slowly Changing Spatial Pattern of the British Columbia Salmon Canning Industry." *Journal of Historical Geography* 14(1) (1988): 22–36.

"The Rationality of Mechanization in the Pacific Salmon-Canning Industry before the Second World War." *Business History Review* 62 (1988): 626–55.

Northcote, T. G. and P. A. Larkin. "The Fraser River: A Major Salmonine Production System." In *Proceedings of the International Large River Symposium*, ed. D. P. Dodge, Canadian Special Publication of Fisheries and Aquatic Sciences 106 (1989), pp. 172–204.

Patrick, K. E. "The Water Resources of British Columbia." In *Inventory of the Natural Resources of British Columbia* (Victoria: British Columbia Natural Resources Conference, 1964): 84–138.

Parry, Mac "The Legitimization of Hell's Gate." *Affairs* 2(24) (1972): 24–9.

Pretious, E. S. "Salmon Catastrophe at Hell's Gate." *BC Professional Engineer* 27 (1976): 13–8.

Ralston, Keith. "Patterns of Trade and Investment on the Pacific Coast, 1867–1892: The Case of the British Columbia Canning Industry." *BC Studies* 1 (1968–1969): 37–45.

Rankin, Murray and Arvay Finlay. "Alcan's Kemano Completion Project: Options and Recommendations." Victoria, BC: Kemano Steering Committee, 1992.

Reid, David J. "Company Mergers in the Fraser River Salmon Canning Industry, 1885–1902." *Canadian Historical Review* 58(3) (1975): 282–302.

Reuss, Martin. "The Art of Scientific Precision: River Research in the United States Army Corps of Engineers." *Technology and Culture* 40(4) (1999): 292–323.

Robinson, J. Lewis. "Fraser River," and "Fraser River Canyon." In *Canadian Encyclopedia*, 2nd ed. Edmonton, Alberta: Hurtig, 1988, pp. 915–6.

Romanoff, Steven. "Fraser Lillooet Salmon Fishing." In *A Complex Culture of the British Columbia Plateau: Traditional Stl'atl'imx Resource Use*, ed. Bryan Hayden. Vancouver: UBC Press, 1992, pp. 222–65.

Roy, Patricia. "The British Columbia Electric Railway Company, 1897–1928: A British Company in British Columbia." PhD thesis, University of British Columbia, Vancouver, BC, 1970.

"Direct Management from Abroad: The Formative Years of the British Columbia Electric Railway." *Business History Review* 47 (1974): 239–59.

"The British Columbia Electric Railway and Its Street Railway Employees: Paternalism in Labour Relations." BC Studies 16 (1973): 3–24.

"The Fine Arts of Lobbying and Persuading: The Case of the BC Electric Railway." In *Canadian Business History: Selected Studies, 1497–1971*, ed. David S. Macmillan. Toronto: McClelland and Stewart, 1972, pp. 239–54.

"The Illumination of Victoria: Late Nineteenth Century Technology and Municipal Enterprise." *BC Studies* 32 (1972–3): 3–24.

Sandwell, R. W. "Finding Rural British Columbia." In *Beyond the City Limits: Rural History in British Columbia*, ed. R. W. Sandwell. Vancouver: UBC Press, 1999, pp. 3–14.

Sewell, W. R. Derrick. "Changing Approaches to Water Management in the Fraser River Basin." In *Environmental Effects of Complex River Development*, ed. Gilbert F. White. Boulder, CO: Westview, 1977, pp. 97–121.

Sparks, W. H. "The Early British Columbia Water Surveys of F. W. Knewstubb." In *Transactions of the Seventh British Columbia Natural Resources Conference*. Victoria: BC Natural Resources Conference, 1954, pp. 29–32.

Stadfeld, Bruce. "Electric Space: Social and Natural Transformations in British Columbia's Hydroelectric Industry to World War II." PhD thesis, University of Manitoba, Winnipeg, MB, 2002.

Taylor, Joseph E., III. "Making Salmon: The Political Economy of Fishery Science and the Road Not Taken." *Journal of the History of Biology* 31 (1998): 33–59.

"Burning the Candle at Both Ends: Historicizing Overfishing in Oregon's Nineteenth-Century Salmon Fisheries." *Environmental History* 4(1) (1999): 54–79.

"The Historical Roots of Canadian-American Salmon Wars." In *Parallel Destinies: Canadian–American Relations West of the Rockies*, ed. John Findlay and Ken Coates. Seattle: University of Washington Press, Montreal/Kingston, Ontario: McGill–Queen's University Press, 2002, pp. 155–80.

Taylor, Mary Doreen. "Development of the Electricity Industry in British Columbia." MA thesis, University of British Columbia, Vancouver, BC, 1965.

Tyrrell, Ian. "American Exceptionalism in an Age of International History." *The American Historical Review* Vol. 96(4) (October 1991): 1031–55.

"Making Nations/Making States: American Historians in the Context of Empire." *Journal of American History* 86(3) (1999): 1015–44.

Wedley, John R. "Infrastructure and Resources: Governments and Their Promotion of Northern Development in British Columbia, 1945–1975." PhD thesis, University of Western Ontario, London, ON, 1986.

"A Development Tool: W. A. C. Bennett and the P. G. E. Railway." *BC Studies* 117 (1998): 29–50.

"Laying the Golden Egg: The Coalition Government's Role in Post-War Northern Development." *BC Studies* 88 (1990–1): 58–92.

"The Wenner–Gren and Peace River Power Development Programs." In *SA TS'E" Historical Perspectives on Northern British Columbia*, ed. Thomas Thorner. Prince George, BC: College of New Caledonia, 1989, pp. 515–545.

White, Gilbert F. "Comparative Analysis of Complex River Development." In *Environmental Effects of Complex River Development*, ed. Gilbert F. White. Boulder, CO: Westview, 1977, pp. 1–22.

White, Richard. "Environmental History, Ecology, and Meaning." *Journal of American History* 76(4) (1990): 1111–6.

"The Nationalization of Nature." *Journal of American History* 86(3) (1999): 976–86.

"Environmental History: Watching a Historical Field Mature," *Pacific Historical Review* 70(1) (2001): 103–11.

Wickwire, Wendy. "'We Shall Drink from the Stream and So Shall You': James A Teit and Native Resistance In British Columbia, 1908–1922." *Canadian Historical Review* 79(2) (1998): 119–236.

Wittner, Shirley. "Barriere: Powerhouse of the Thompson." In *Reflections: Thompson Valley Histories*, ed. Wayne Norton and Wilf Schmidt. Kamloops, BC: Plateau Press, 1994, pp. 152–7.

Woo, Ming-Ko. "Water in Canada, Water for Canada." *Canadian Geographer* 45(1) (2001): 85–92.

Worster, Donald. "Doing Environmental History." In *Ends of the Earth: Perspectives on Modern Environmental History*, ed. Donald Worster. Cambridge, England: Cambridge University Press, 1988, pp. 289–307.

"Transformations of the Earth: Toward an Agroecological Perspective in History." *Journal of American History* 76(4) (1990): 1087–106.

Wyatt, David. "The Thompson." In *Plateau*, Vol. 12 of *Handbook of North American Indians*, ed. Deward E. Walker Jr. Washington, D.C.: Smithsonian Institution, 1998, pp. 191–202.

Wynn, Graeme. "The Rise of Vancouver." In *Vancouver and Its Region*, ed. Graeme Wynn and Timothy Oke. Vancouver: UBC Press, 1992, pp. 69–145.

3. Books

Fraser River Pile Driving: The Company History. New Westminster, BC: Camart Studio, 1976.

Looking Forward, Looking Back. Vol. I. *Report of the Royal Commission on Aboriginal Peoples.* Ottawa: Minister of Supply and Services Canada, 1996.

Allard, Dean Conrad, Jr. *Spencer Fullerton Baird and the U.S. Fish Commission.* New York: Arno, 1978.

Armstrong, Christopher. *The Politics of Federalism: Ontario's Relations with the Federal Government, 1867–1942.* Toronto: University of Toronto Press, 1981.

Armstrong, Christopher and H. V. Nelles. *Monopoly's Moment: The Organization and Regulation of Canadian Utilities, 1830–1930.* Philadelphia: Temple University Press, 1986.

Barman, Jean. *The West Beyond the West: A History of British Columbia.* Toronto: University of Toronto Press, 1991.

Biagloli, Mario, ed. *The Science Studies Reader.* New York/London: Routledge, 1999.

Biswas, Asit K. *History of Hydrology.* Amsterdam/London: North-Holland, 1970.

Blatter, Joachim and Helen Ingram, eds. *Reflections on Water: New Approaches to Transboundary Conflicts and Cooperation.* Cambridge, MA: MIT Press, 2001.

Bocking, Richard. *Mighty River: A Portrait of the Fraser.* Vancouver: Douglas & McIntyre, 1997.

Bocking, Stephen. *Ecologists and Environmental Politics: A History of Contemporary Ecology.* New Haven, CT: Yale University Press, 1997.

Bothwell, Robert and William Kilbourn. *CD Howe: A Biography.* Toronto: McClelland and Stewart, 1979.

Brigham, Jay L. *Empowering the West: Electrical Politics Before FDR.* Lawrence: University of Kansas Press, 1998.

Brown, Robert Craig and Ramsay Cook. *Canada 1896–1921: A Nation Transformed.* Toronto: McClelland and Stewart, 1974.

Cail, Robert E. *Land, Man and the Law: The Dispersal of Crown Lands in British Columbia, 1871–1913.* Vancouver: UBC Press, 1974.

Cameron, Laura. *Openings: A Meditation on History, Method and Sumas Lake.* Montreal: McGill–Queen's University Press, 1997.

Campbell, Duncan C. *Global Mission: The Story of Alcan.* 3 vols. Montreal: Alcan, 1985–1990.

Carlson, Keith, ed. *A Sto:lo Coast Salish Historical Atlas.* Vancouver: Douglas & McIntyre, 2001.

Carrothers, W. A. *The British Columbia Fisheries.* Toronto: University of Toronto Press, 1941.

Cholderhose, R. J. and Marj Trim. *Pacific Salmon and Steelhead Trout.* Vancouver: Douglas & McIntyre, 1979.

Christensen, Bev. *Too Good to Be True: Alcan's Kemano Completion Project.* Vancouver: Talonbooks, 1995.

Clemens, W. A. *Education and Fish.* Nanaimo, BC: Fisheries Research Board of Canada Station, MS Report Series No. 974, 1968.

Cole, Stephen. *Making Science: Between Nature and Society.* Cambridge, MA: Harvard University Press, 1992.

Crane, Diana. *Invisible Colleges: Diffusion of Knowledge in Scientific Communities.* Chicago: University of Chicago Press, 1972.

Creighton, Donald. *The Commercial Empire of the St. Lawrence.* Rev. ed. Toronto: Macmillan, 1956.

Cronon, William. *Nature's Metropolis: Chicago and the Great West.* New York: Norton, 1991.

Crowcroft, Peter. *Elton's Ecologists: A History of the Bureau of Animal Populations.* Chicago: University of Chicago Press, 1991.

De Nevarre Kennedy, John. *History of the Department of Munitions and Supply Canada in the Second World War.* Ottawa: King's Printer, 1950, Vol. II.

De Solla Price, Derek J. *Little Science, Big Science...And Beyond.* New York: Columbia University Press, 1986 (first published 1968).

Dorsey, Kurkpatrick. *The Dawn of Conservation Diplomacy: US–Canadian Wildlife Protection Treaties in the Progressive Era.* Seattle: University of Washington Press, 1998.

Drake-Terry, Joanne. *The Same as Yesterday: The Lillooet Chronicle the Theft of Their Lands and Resources.* Lillooet, BC: Lillooet Tribal Council, 1989.

Drushka, Ken. *HR: A Biography of HR MacMillan.* Madeira Park, BC: Harbour Publishing, 1995.

Duff, Wilson. *The Indian History of British Columbia: The Impact of the White Man.* Victoria: Royal BC Museum, Memoir No 5, 1969.

Dunlap, Thomas. *Nature and the English Diaspora: Environment and History in the United States, Canada, Australia and New Zealand.* Cambridge, England: Cambridge University Press, 1999.

Ewert, Henry. *The Story of the BC Electric Railway Company.* North Vancouver: Whitecap Books, 1986.

Fleming, Keith R. *Power at Cost: Ontario Hydro and Rural Electrification, 1911–1958*. Montreal/Kingston: McGill–Queen's Press, 1992.

Follansbee, Robert. *A History of the Water Resources Branch, US Geological Survey: Volume 1, From Predecessor Surveys to June 30, 1919*. Washington, D.C.: U.S. Geological Survey, 1994 (first published 1938).

Froschauer, Karl. *White Gold: Hydroelectric Power in Canada*. Vancouver: UBC Press, 1999.

Galison, Peter and Bruce Hevly, eds. *Big Science: The Growth of Large Scale Research*. Stanford, CA: Stanford University Press, 1992.

Girard, Michel F. *L'Ecologisme Retrouvé: Essor at déclin de la Commission de la Conservation du Canada*. Ottawa: University of Ottawa Press, 1994.

Goldsmith, Edward and Nicholas Hildyard, eds. *The Social and Environmental Effects of Large Dams*. Cornwall: Wadebridge Ecological Center, 1984, 1986, Vols. I and II.

Golinski, Jan. *Making Natural Knowledge: Constructivism and the History of Science*. Cambridge, England: Cambridge University Press, 1998.

Grant, Shelagh D. *Sovereignty or Security? Government Policy in the Canadian North 1936–1950*. Vancouver: UBC Press, 1988.

Groot, C. and L. Margolis, eds. *Pacific Salmon Life Histories*. Vancouver, UBC Press, 1991.

Grove, Richard H. *Green Imperialism: Colonial Expansion, Tropical Island Edens and the Origins of Conservation, 1600–1860*. Cambridge, England: Cambridge University Press, 1995.

Harden, Blaine. *A River Lost: The Life and Death of the Columbia*. New York: Norton, 1996.

Harris, Cole. *The Resettlement of British Columbia: Essays on Colonialism and Geographical Change*. Vancouver: UBC Press, 1997.
   Making Native Space: Colonialism, Resistance, and Reserves in British Columbia. Vancouver: UBC Press, 2002.

Harris, Douglas C. *Fish, Law and Colonialism: The Legal Capture of Salmon in British Columbia*. Toronto: University of Toronto Press, 2001.

Harvey, Mark. *A Symbol of Wilderness: Echo Park and the American Conservation Movement*. Seattle: University of Washington Press, 2000.

Headrick, Daniel R. *The Tentacles of Progress: Technology Transfer in the Age of Imperialism, 1850–1940*. New York: Oxford University Press, 1988.

Hessing, Melody and Michael Howlett. *Canadian Natural Resource and Environmental Policy: Political Economy and Public Policy*. Vancouver: UBC Press, 1999.

Hughes, Thomas P. *Networks of Power: Electrification in Western Society*. Baltimore: Johns Hopkins University Press, 1983.

Hume, Mark. *Adams River: The Mystery of the Adams River Sockeye*. Vancouver: New Star Books, 1994.

Hundley, Norris, Jr. *The Great Thirst: Californians and Their Water, 1770s–1990s*. Berkeley: University of California Press, 1992.

Hutchison, Bruce. *The Fraser*. Toronto: Holt Rinehart, 1950.

Canada: Tomorrow's Giant. Toronto: Longmans, Green, 1957.

Innis, Harold Adams. *The Fur Trade in Canada: An Introduction to Canadian Economic History*, rev. ed. Toronto: University of Toronto Press, 1970.

Essays in Canadian Economic History, ed. Mary Quayle Innis. Toronto: University of Toronto Press, 1956.

Staples, Markets and Cultural Change, ed. Daniel Drache. Montreal: McGill–Queen's Press, 1995.

Jackson, Donald C. *Building the Ultimate Dam: John S. Eastwood and the Control of Water in the West*. Lawrence: University Press of Kansas, 1995.

Johnstone, Kenneth. *The Aquatic Explorers: A History of the Fisheries Research Board of Canada*. Toronto: University of Toronto Press, 1977.

Kahrer, Gabrielle. *From Speculative to Spectacular: The Seymour River Valley 1870s to 1980s: A History of Resource Use*. Vancouver: Greater Vancouver Regional District Parks, 1989.

Kendrick, John. *People of the Snow: The Story of Kitimat*. Toronto: NC Press, 1987.

Kohler, Robert E. *Partners in Science: Foundations and Natural Scientists 1900–1945*. Chicago: University of Chicago Press, 1991.

Laforet, Andrea and Annie York. *Spuzzum: Fraser Canyon Histories, 1808–1939*.Vancouver: UBC Press, 1998.

Lamb, W. Kaye, ed. *The Letters and Journals of Simon Fraser, 1806–1808*. Toronto: Macmillan, 1960.

Langston, Nancy. *Forest Dreams, Forest Nightmares: the Paradox of Old Growth in the Inland West*. Seattle: University of Washington Press, 1995.

Latour, Bruno. *We Have Never Been Modern*. Cambridge, MA: Harvard University Press, 1993.

Pandora's Hope: Essays on the Reality of Science Studies. Cambridge, MA: Harvard University Press, 1999.

Logan, Harry T. *Tuum Est: A History of the University of British Columbia*. Vancouver: UBC Press, 1958.

Lone, Joseph and Sandy Ridlington, eds. *The Northwest Salmon Crisis: A Documentary History*. Corvallis: Oregon University Press, 1996.

Loo, Tina. *Making Law, Order and Authority in British Columbia, 1821–1871*. Toronto: University of Toronto Press, 1994.

Lyons, Cicely. *Salmon Our Heritage: The Story of a Province and an Industry*. Vancouver: BC Packers, 1969.

Mackie, Richard Somerset. *Trading Beyond the Mountains: The British Fur Trade on the Pacific, 1793–1843*. Vancouver: UBC Press, 1997.

Manore, Jean L. *Cross-Currents: Hydro-Electricity and the Engineering of Northern Ontario*. Waterloo, Ontario: Wilfred Laurier University Press, 1999.

McCully, Patrick. *Silenced Rivers: The Ecology and Politics of Large Dams*. London: Zed Books, 1996.

McCutcheon, Sean. *Electric Rivers: The Story of the James Bay Project*. Montreal: Black Rose Press, 1991.

McDonald, Robert A. J. *Making Vancouver: Class, Status, and Social Boundaries, 1863–1913*. Vancouver: UBC Press, 1996.

McEvoy, Arthur F. *The Fisherman's Problem: Ecology and Law in the California Fisheries, 1850–1980*. Cambridge, England: Cambridge University Press, 1986.

McNeill, John R. *Something New Under the Sun: An Environmental History of the Twentieth-Century World*. New York: Norton, 2000.

Meggs, Geoff. *Salmon: The Decline of the British Columbia Fishery*. Vancouver: Douglas & McIntyre, 1991.

Metcalfe, E. Bennett. *A Man of Some Importance: The Life of Roderick Langmere Haig-Brown*. Seattle/Vancouver: Wood Publishers, 1985.

Mighetto, Lisa and Wesley J. Ebel. *Saving the Salmon: A History of the US Army Corps of Engineers' Efforts to Protect Anadromous Fish on the Columbia and Snake Rivers*. Seattle: Historical Research Associates, 1994.

Merchant, Carolyn, ed. *Green Versus Gold: Sources in California Environmental History*. Washington, D.C.: Island Press, 1998.

Mitchell, David. *WAC Bennett and the Rise of British Columbia*. Vancouver: Douglas & McIntyre, 1983.

Morley, Alan. *Vancouver: From Milltown to Metropolis*. Vancouver: Mitchell Press, 1961.

Mouat, Jeremy. *The Business of Power: Hydro-Electricity in Southeastern British Columbia, 1897–1997*. Victoria: Sono Nis Press, 1997.

Mount, Jeffrey F. *California Rivers and Streams: The Conflict Between Fluvial Process and Land Use*. Berkeley: University of California Press, 1995.

Muszynski, Alicja. *Cheap Wage Labour: Race and Gender in the Fisheries of British Columbia*. Montreal/Kingston: McGill–Queen's University Press, 1996.

Nash, Gerald D. *World War II and the West: Reshaping the Economy*. Lincoln: University of Nebraska Press, 1990.

Nelles, H. V. *The Politics of Development: Forests, Mines and Hydro-Electricity in Ontario, 1849–1941*. Toronto: Macmillan, 1974.

Neuberger, Richard L. *Our Promised Land*, ed. David L Nicandri. Moscow: University of Idaho Press, 1989 (first published 1938).

New, W. H. *Land Sliding: Imagining Space, Presence, and Power in Canadian Writing*. Toronto: University of Toronto Press, 1997.

Newell, Dianne. *Tangled Webs of History: Indians and the Law in Canada's Pacific Coast Fisheries*. Toronto: University of Toronto Press, 1993.

— ed. *The Development of the Pacific Salmon-Canning Industry: A Grown Man's Game*. Montreal: McGill–Queen's University Press, 1989.

Newell, Dianne and Rosemary E. Ommer, eds. *Fishing Places, Fishing Peoples: Traditions and Issues in Canadian Small-Scale Fisheries*. Toronto: University of Toronto Press, 1999.

Norrie, Kenneth and Douglas Owram. *A History of the Canadian Economy.* Toronto: Harcourt Brace Jovanovich, 1991.

Nye, David E. *Electrifying America: Social Meanings of a New Technology, 1880–1940.* Cambridge, MA: MIT Press, 1990.

Pethic, Derek. *British Columbia Disasters.* Langley, BC: Stagecoach Publishing, 1978.

Pisani, Donald J. *To Reclaim a Divided West: Water, Law and Public Policy, 1848–1902.* Albuquerque: University of New Mexico Press, 1992.

Pitzer, Paul C. *Grand Coulee: Harnessing a Dream.* Pullman: Washington State University Press, 1994.

Polanyi, Karl. *The Livelihood of Man,* ed. Harry W Pearson. New York: Academic, 1977.

Rajala, Richard. *Clearcutting the Pacific Rain Forest: Production, Science, and Regulation.* Vancouver: UBC Press, 1998.

Ray, Arthur J. *Indians in the Fur Trade.* Toronto: University of Toronto Press, 1974.

   I Have Lived Here Since the World Began. Toronto: Lester Publishing and Key Porter Books, 1996.

Regehr, T. D. *The Canadian Northern Railway: Pioneer Road of the Northern Prairies, 1895–1918.* Toronto: Macmillan, 1976.

Robin, Martin. *The Rush for Spoils: The Company Province, 1871–1933.* Toronto: McClelland and Stewart, 1972.

   Pillars of Profit: The Company Province, 1934–1972. Toronto: McClelland and Stewart, 1973.

Roos, John F. *Restoring Fraser River Salmon: A History of the International Pacific Salmon Fisheries Commission, 1937–1985.* Vancouver: Pacific Salmon Commission, 1991.

Rounsefell, George. *Ecology, Utilization, and Management of Marine Fisheries.* St. Louis, MO: Mosby, 1975.

Roy, Patricia. *Vancouver: An Illustrated History.* Toronto: Lorimer, 1980.

Scarce, Rik. *Fishy Business: Salmon, Biology, and the Social Construction of Nature.* Philadelphia: Temple University Press, 2000.

Scott, Anthony and Philip A. Neher. *The Public Regulation of Commercial Fisheries in Canada.* Ottawa: Economic Council of Canada, 1981.

Sewell, W. R. Derrick. *Water Management and Floods in the Fraser Basin.* Chicago: Department of Geography, Research Paper No. 100, 1965.

Shallat, Todd. *Structures in the Stream: Water, Science, and the Rise of the US Army Corps of Engineers.* Austin: University of Texas Press, 1994.

Shapin, Stephen and Simon Schaffer. *Leviathan and the Air Pump: Hobbes, Boyle, and the Experimental Life.* Princeton, NJ: Princeton University Press, 1985.

Shapin, Stephen. *A Social History of Truth: Civility and Science in Seventeenth-Century England.* Chicago: University of Chicago Press, 1994.

Sheets-Pyenson, Susan. *Cathedrals of Science: The Development of Colonial Natural History Museums During the Late Nineteenth Century*. Montreal: McGill–Queen's University Press, 1988.

Sherman, Paddy. *Bennett*. Toronto: McClelland and Stewart, 1966.

Shrum, Gordon with Peter Stursberg. *Gordon Shrum: An Autobiography*, ed. Clive Cocking. Vancouver: UBC Press, 1986.

Smith, Courtland L. *Salmon Fishers of the Columbia*. Corvallis: Oregon State University Press, 1979.

Smith, George David. *From Monopoly to Competition: The Transformation of Alcoa, 1888–1986*. New York: Cambridge University Press, 1988.

Smith, Tim. *Scaling Fisheries: The Science of Measuring the Effects of Fishing, 1855–1955*. Cambridge, England: Cambridge University Press, 1994.

Stacey, C. P. *Canada and the Age of Conflict Volume 2: The Mackenzie King Era*. Toronto: University of Toronto Press, 1981.

Swainson, Neil. *Conflict Over the Columbia: The Canadian Background to an Historic Treaty*. Montreal/Kingston: McGill–Queen's University Press, 1979.

Swettenham, John. *McNaughton, Vol. 3. 1944–1966*. Toronto: Ryerson, 1969.

Taylor, Joseph E., III. *Making Salmon: An Environmental History of the Northwest Fisheries Crisis*. Seattle: University of Washington Press, 1999.

Teit, James Alexander. *The Thompson Indians of British Columbia*. The Jesup North Pacific Expedition. Vol. 2, part 4. American Museum of Natural History Memoir 2. New York: Knickerbocker Press.

Tennant, Paul. *Aboriginal People and Politics: The Indian Land Question in British Columbia, 1849–1989*. Vancouver: UBC Press, 1990.

Thompson, John Herd and Stephen J. Randall. *Canada and the United States: Ambivalent Allies*. Montreal/Kingston: McGill-Queen's University Press, 1994.

Thompson, John Herd with Alan Seager. *Canada 1922–1939: Decades of Discord*. Toronto: McClelland and Stewart, 1986.

Tobey, Ronald C. *Technology as Freedom: The New Deal and the Electrical Modernization of the American Home*. Berkeley: University of California Press, 1996.

Tomasevich, Jozo. *International Agreements on Conservation of Marine Resources with Special Reference to the North Pacific*. Stanford, CA: Food Research Institute, 1943.

Tyrrell, Ian. *True Gardens of the Gods: Californian–Australian Environmental Reform, 1860–1930*. Berkeley: University of California Press, 1999.

Waldram, James B. *As Long as the Rivers Run: Hydroelectric Development and Native Communities in Western Canada*. Winnipeg: University of Manitoba Press, 1988.

Wheelwright, Philip. *Heraclitus*. New York: Athenum, 1964.

White, Richard. *Land Use, Environment, and Social Change: The Shaping of Island County, Washington*. Seattle: University of Washington Press, 1980.

The Organic Machine: The Remaking of the Columbia River. New York: Hill and Wang, 1995.

Williston, Eileen and Betty Keller. *Forests, Power and Policy: The Legacy of Ray Williston*. Prince George, BC: Caitlin Press, 1997.

Wilson, Jeremy. *Talk and Log: Wilderness Politics in British Columbia, 1965–1996*. Vancouver: UBC Press, 1998.

Worster, Donald. *Dust Bowl: The Southern Plains in the 1930s*. New York: Oxford University Press, 1979.

Rivers of Empire: Water, Growth, and the American West. New York: Pantheon, 1985.

ed. *The Ends of the Earth: Perspectives on Modern Environmental History*. New York: Cambridge University Press, 1988.

Under Western Skies: Nature and History in the American West. New York: Oxford University Press, 1992.

An Unsettled Country: Changing Landscapes of the American West. Albuquerque: University of New Mexico Press, 1994.

Wynn, Graeme. *Timber Colony: A Historical Geography of Early Nineteenth Century New Brunswick*. Toronto: University of Toronto Press, 1981.

Zaslow, Morris. *Reading the Rocks: The Story of the Geological Survey of Canada, 1842–1972*. Ottawa: Macmillan, 1975.

Zeller, Suzanne. *Inventing Canada: Early Victorian Science and the Dream of a Transcontinental Nation*. Toronto: University of Toronto Press, 1987.

Index

Adams, E. H., 132
Adams River, 34, 71, 104
Aggasiz, 215
Alaska, 154, 156, 166
Alberta, 54, 154
Alcoa, 152, 155
Alexander, George, 160
Allen, E. W., 87
Alouette Lake, 57, 64, 77
Alouette River, 57, 64, 77
Aluminum, 149–178, 180, 270
Aluminum Company of Canada
    (Alcan), 15, 17, 150, 184, 185,
    193, 202, 205, 232, 271, 272,
    275
  Expansion of during the Second
    World War, 152
  Map of northern BC project,
    171
  Project development in BC,
    153–157
Amur River, 268
Anderson Creek, 95
Andrews, F. J., 261
Anscomb, Herbert, 119, 145
Arrow Lakes, 142
Arvida, 152

Babcock, John Pease, 34, 115
  Dams, remedial actions on, 71, 75,
    77

Fisheries conditions, views
    concerning, 20, 48, 49, 70, 104
  Hells Gate slides' effects,
    descriptions of, 19, 27, 28, 35, 98
  Salmon at Hells Gate, report on,
    28–31
  Salmon migrations, views
    concerning, 28
Babine Lake, 38
Barrett, Dave, 228
Barriere River, 64
Bell, Milo, 185, 236
  Alcan development, concerns
    regarding, 164
  Fishways at Hells Gate, design of,
    101, 102, 236
Bell-Irving, Henry, 47
Bennett, W. A. C., 122, 198, 223,
    227–228, 263
  Peace River development,
    announcement of, 179–182
Biological Board of Canada, 92, 108,
    111
Black, Edgar, 251
Board of Engineers, 92
Bonner, Robert, 200
Bonneville Power Administration,
    119, 130, 134, 139, 140, 147,
    152, 273
Boston Bar, 24, 40
Bow River, 54

Brennan, B. M., 87, 88
Brett, J. R., 236, 237, 241, 242, 243, 249, 251
Bridge River, 115, 139, 189, 199, 235, 269
  Delayed development of, 66, 130, 134, 213
  Development of, 139–140, 202
  Development, environmental and social effects of, 140–141, 147
  Development proposals for during the Second World War, 131, 136
  Early development plans for, 64, 66, 119
  Inauguration of power project on, 119, 121, 148
BC Canner's Association, 43
BC Electric, 15, 119–120, 146, 162, 178, 184, 204, 206, 211, 222, 260, 263
  and the Bridge River project, 119–120, 130, 213, 269
  and the Columbia-to-Fraser diversion, 199
  and the Moran project, 193, 194, 202
  and Thermal power, 222, 223
  and Transmission costs, 192
  Business and politics of during the Second World War, 125, 126, 129, 130, 131, 132–134, 146–147
  Fish-Power research, sponsorship of, 216, 219, 225, 240, 243, 248, 256, 263
  Name changes of, 18
  Nationalization of, 182
  see also B. C. Electric Railway Company
BC Electric Railway Company (BCER), 56, 58, 59, 60–62, 64, 66–67, 68, 69, 72–73, 74, 75, 77, 82, 83, 128
  BC market, entrance to, 58
  Corporate name changes of, 17–18
  Expansion of (1920s), 63–64
  Power production (1905–1928), graph of, 65
  WCPC, cooperation with, 62–63
  see also BC Electric
BC Energy Board, 209, 220, 226, 262
BC Federation of Labour, 217
BC Fish and Game Branch, 232, 238, 239, 243, 244, 246, 247
BC Fish and Wildlife Branch, 227
BC Game Commission, 251
BC Hydro, 122, 182, 227, 229, 270
BC Natural Resources Conference, 184, 195, 208, 209, 236
BC Packers, 45, 159, 169, 239
BC Power Commission (BCPC), 138, 139, 155, 184, 204, 206, 211, 217, 247, 260
  Establishment and mandate of, 137
  Fraser Basin, investigations of, 202
  System, building of, 139, 141, 148, 178
BC Water Branch, 117, 186
British North America Act (1867), 8
Brown, Joe, 95
Buchanan, J. M., 169, 239, 240
Buntzen, Johannes, 74
Buntzen Lake, 57, 60, 61
Burdis, W. D., 70–73
Burrard Inlet, 57, 60, 133
Bute Inlet, 151

California Fish and Game Commission, 27
Campbell River, 142, 147
Canadian Commonwealth Federation (CCF) (provincial), 120, 122, 127, 134, 138, 165
Canadian Fisheries Research Board, 109, 110–112
Canadian Fisherman's Weekly, 110

Canadian Fishing Company, 161, 168
Canadian National Railways, 144, 177
Canadian Northern Railway (CNR), 24, 29, 32, 35, 103
Canadian Pacific Railway (CPR), 8, 9, 24, 31, 34, 41, 144
  Confederation and, 8
  Hells Gate slides, employees' views of, 29, 30, 31, 34
Canol pipeline, 154
Cariboo Hydraulic Mining Company, 156
Cariboo River, 223
Carrothers, W. A., 128
Carson, E. H., 194
Cheslatta T'en, 15, 24, 175–176
Cheslatta dam, 175, 205
Cheslatta Lake, 171, 174
Chilcotin River, 30
Chilko Lake,
  Fisheries defense of, 158, 164, 168, 169, 172, 271, 272
  Fisheries surveys of, 142, 163
  Power surveys of, 150–151, 155, 202
China Bar, 25, 30, 35
Clark, George, 161–163
Clay, C. H., 255
Clayquot Sound, 38
Clearwater River, 202, 223
Clemens, W. A.,
  Fish-power research, concerns regarding, 250, 251, 252
  Tagging studies at Hells Gate, proposals for, 50, 92
  W. F. Thompson, criticisms of, 110, 111
Coalition (provincial government), 124, 126–128, 155, 167
Collins, G., 251
Colorado River, 54
Columbia River, 2, 7, 12, 16, 17, 21, 81, 122, 124, 146, 147, 153, 192, 222, 262, 267, 270, 273

Coyote and, 23
Fish-Power research on, 241, 249, 264
Flooding of, 144
Fraser River, as a contrast to, 157, 215, 235, 262, 269, 272–273
Upper basin projects, fishing interests' support of, 217–218
Upper basin projects, map of, 181
Upper basin projects, politics of, 16, 142, 180, 182, 183, 195–203, 205, 213, 222, 228–229, 263, 274, 275
  see also Columbia River dams and Columbia-to-Fraser diversion
Columbia River dams, 54, 77–78, 80, 100, 101, 125, 152, 214, 215, 271
  Fish passage at, 100, 101, 236, 249, 251
  see also Columbia River
Columbia River Treaty, 3, 180, 182, 214, 217, 229, 275
Columbia-to-Fraser diversion, 183, 220, 222, 226, 228, 259, 270, 272, 274
  Politics and promotion of, 195–203, 223
  Public reception of, 206, 213, 214
  see also Columbia River
Commission of Conservation, 58, 78
Conservative Party (federal), 80, 182, 222
Conservative Party (provincial), 122, 128, 145
Coquitlam Lake, 57, 60, 61, 74, 75, 76
Coquitlam River, 57, 59, 60, 72–73, 76
Coquitlam–Buntzen project, 57, 61, 62, 64, 73
Cornett, J. W., 135, 138
Cowichan River, 38

Coyote, 22–23, 27, 52
Crippen Wright Engineering, 189
Cultus Lake, 89, 90, 105, 109
Cunningham, F. H., 32, 37, 41, 43

Dane, Albert E., 259
Davis, Ernest, 123, 124, 128, 129,
    154
Devlin, A. G., 257, 258
Department of External Affairs,
    165
Department of Fisheries (federal), 254,
    259
  and the Fraser Basin Board, 221,
    223
  Alcan project, regulation of, 165,
    170–172, 173–175
  Fisheries defense, role in, 160,
    161, 203, 204, 245, 271, 272,
    273
  Fisheries research, sponsorship of,
    90, 157, 158, 163, 238, 243, 244,
    246–247, 248, 251, 255, 264,
    265
  Legal authority of, 159, 160, 264,
    271
  Name changes of, 18
  see also Department of Marine and
    Fisheries
Department of Fisheries (provincial),
    30, 160, 161, 162, 238
Department of Indian Affairs, 36,
    175
Department of the Interior, 78
Department of Justice, 160
Department of Marine and Fisheries,
  and the Hells Gate slides, 30, 31,
    38, 39, 48, 50
  and Native fishing, 36, 37, 38, 41,
    42, 76
  Tributary dams, actions regarding,
    70, 73, 75, 77
  See also Department of Fisheries
    (federal)
Department of Munitions and
    Supplies, 131

Department of Northern Affairs and
    National Resources, 204, 205
Department of Trade and Commerce,
    188, 189
Deutsh, John, 260
Diefenbaker, John, 218
Douglas Channel, 171, 176
Downtown, Geoffrey, 119
Doyle, Henry, 12, 226
Dubose, McNeely, 164, 166, 169,
    173–174, 185, 235
Duncan, 121, 126
Dynesius, Mats, 1, 267

Eagle Pass, 197
Eckman, James, 168–169
Eggers, Hans, 193, 194
Eutsuk Lake, 171, 175

Farewell Canyon, 160, 235
Farrow, Richard, 150, 170
Final Report of the Board of Engineers
    (1928), 49
Finn, Donald, 260
Fisheries Act (federal), 161, 162,
    264
Fisheries Act (provincial), 38, 71, 159,
    160, 170
Fisheries Association of BC, 185, 203,
    204, 208, 240, 260
Fisheries Development Council, 244,
    245, 247, 255
Fisheries Protection Committee, 217,
    226
Fisheries Research Board of Canada
    (FRBC), 163, 236, 238, 239, 243,
    244, 246, 247, 251, 252
Fisheries Science, 16, 27, 48–51
  Conducted at Hells Gate, *see* Hells
    Gate
  Fish-power research, politics of,
    231–266
Fishing, 46
  and the Fish vs. Power debate, *see*
    Hydro electricity
  and the search for a treaty, 84–86

Hells Gate slides, effects of, 43–48
Native fishery, 36–43
Native fishing stations, 21
Sockeye catches (1901–1933), 46
Tags and, 93–97
Tributary dams, effects of, 69–71,
    83
Fishways, 71–72, 73–75, 84,
    100–102, 115, 236, 240, 251,
    257
Foerster, Russell, 89, 105, 111
Found, W. A., 87
Frame, S. H., 186
Fraser, Simon, 7, 23
Fraser Basin Board, 190, 192, 204,
    264
    Criticisms of, 225, 226
    Dam development program, 148,
        183, 194, 202, 216, 220–221,
        223, 225
    Formation of, 145
    Site investigations, map of, 224
Fraser Canyon, 7, 20, 22, 36, 39, 40,
    42, 52, 53, 55, 59, 64, 66, 72, 90,
    95, 103, 139, 143, 151, 256
    and the Hells Gate slides, *see* Hells
        Gate
    Coyote and, 23
    Early description of, 23–24
    Map of ca. 1916, 26
    Native occupation of, 21
Fraser River, 2, 16, 17, 19, 20, 21, 23,
    24, 27
    and Alcan, *see* Alcan
    and Fisheries Science, *see* Fisheries
        Science
    and Hells Gate investigations, *see*
        Hells Gate
    and Hells Gate slides, *see* Hells Gate
    Colonization and development of,
        7–10
    Conditions during Hells Gate slides,
        see Hells Gate
    Debates over development of, *see*
        Hydro electricity
    Early power surveys of, 59

Floods, *see* Fraser River flood
    (1948)
Map of, 6
Native fishery on, *see* Fishing
Perceptions of as a power source in
    the Second World War,
    123–124
Physical characteristics of, 5
Present and future of, 13
Tributary development of, 53–83,
    see also Hydro electricity
Fraser River Canners' Association, 70,
    73
Fraser River flood (1948), 121,
    142–146
    Map of, 143
Fraser River Multiple Use
    Committee, 204, 218, 222, 223,
    253, 256
Freeman, Miller, 99
Fry, F. E., 251

Gaglardi, Phil, 227
Geen, G. H., 261
Gilbert, Charles, 28, 34, 89, 90
Gold Rush (1858), 7, 24, 53
Grand Coulee dam, *see* Columbia
    River dams
Grauer, Dal, 199, 211, 243
Gruening, Ernest, 156
Gwyther, Val, 235, 253–255

Hager, A. L., 87
Haig-Brown, Roderick, 185, 207, 211,
    234
Harris, Charles W., 101
Hart, John, 126–127, 138, 142,
    147
Hawley-Smoot tariff, 67
Heena, Paul, 40
Hells Gate, 13, 53, 55, 76, 82, 114,
    157–158, 232, 233, 234–235,
    236, 237, 240, 250, 256, 271,
    274
    As a native fishing site, 21–22
    Coyote and, 22–23

Hells Gate (*cont.*)
  Difficulties of water transport
    through, 24
  Landslide effects, map of , 25
  Landslides and aftermath at, 19–52
  Location and characteristics of, 20
  Photographs of, 21, 22, 27, 33, 85,
    92
  Scientific examination and
    reconstruction of, 84–117, 158
History of Science, 12
Hitner, Walter, 101
Hoar, W. S., 242–243, 244, 251, 265
Homathko River, 151
Hoover dam, 80, 214
Hope, 37, 57, 59, 95, 104, 143
Horsefly River, 70, 72
Horston, W. R., 251
Howard, G. V., 97
Howe, C. D., 133, 167
Huber, Bill, 150
Hudson's Bay Company (HBC), 23,
    24
Hutchison, Bruce, 185, 207
Hydro electricity, 12, 179–229
  Early development in BC of, 56–69
  Post-war developments of, 149–178
  Second World War, state of during,
    122–142

Idaho, 202
Indian Rights Association, 39
Industrial Development Act (IDA),
    162–163
Ingledow, Tom, 199
Innis, Harold, 10, 11
Institute of Fisheries, see University of
    British Columbia (UBC)
Interior and Insular Affairs Committee
    (U.S. Senate), 201
Interior Tribes, 39
International Convention of High Seas
    Fisheries of the North Pacific
    Ocean, 265
International Council for the
    Exploration of the Sea, 89

International Engineering Company,
    150
International Joint Commission (IJC),
    183, 201, 202, 204, 217, 222,
    270
  Columbia River investigations, role
    in, 195–197
International Pacific Salmon Fisheries
    Commission (IPSFC), 84, 86, 89,
    102, 105, 111, 112, 115, 116,
    158, 160, 185, 207, 209, 225,
    234–235, 238, 239, 249, 254,
    256, 258, 261
  Columbia-to-Fraser diversion,
    concerns regarding, 200
  Criticisms of, 108, 109–110,
    111–112, 116
  Dam projects, scientific
    investigations of, 163–165, 169,
    172, 195, 216, 243, 264
  Establishment of, 14, 86
  Fisheries defense, role in, 203, 204,
    221, 236
  Organization of, 87
  Research Mandate and Approach
    of, 89, 90, 91, 107–108, 239,
    244, 246, 247, 248
International Salmon Investigation
    Federation, 82
International Waterways Treaty,
    201
International Woodworkers of
    America (IWA), 217

Jackson, Charles, 87
Johnnie, Chief, 73
Johnson, Byron, 164, 165, 166, 185,
    207
Jordan, David Starr, 28
Jordan, Leonard, 202
Juan de Fuca Strait, 43

Kaiser Corporation, 198, 199, 200,
    213
  Columbia River, proposal for
    development of, 180, 196, 197

Kamloops, 22, 64, 125, 141, 212
Kania, J. E., 218
Kask, Jack L., 94–96, 112
Kemano, 17, 171
Kemano River, 176
Kenney Dam, 171, 172
Kenney, E. T., 122, 150, 156, 157, 161, 166, 169
Khatanga River, 268
Kiernan, Kenneth, 219
Kitimat, 171, 176, 186
Kohler, Robert E., 266
Korean War, 167–168

Langston, Nancy, 12
Larkin, Peter, 225
  Career of, 238
  Fish-power research, perspectives on, 231–232, 242, 243, 248, 255, 262, 265
Laurier, Wilfred, 38
Lena River, 268
Lesage, Jean, 198, 222
Liberal Party (federal), 158, 180
Liberal Party (provincial), 122, 128, 200
Lillooet, 186
Lower Mainland, 190, 192
Lulu Island, 45
Lytton, 8, 37, 41, 95, 183, 199

McBride, Richard, 58, 71
Macdonald, J. C., 128
McGregor River, 223
McHugh, J. H., 31, 32, 36, 48
  Hells Gate landslides, assessment of, 32, 35
  Native fishers, view of, 36, 41
McKenna–McBride Commission, 39
Mackenzie, Norman, 256
Mackenzie River, 2, 267
MacMillan, H. R., 159, 225, 239–240, 248

McNaughton, Andrew, 183, 196–202, 210, 213, 214, 217, 218, 220, 222, 223, 228, 259
  Canadian development of the Columbia River, advocacy for, 196–202, 213, 215, 223, 270
Mainwaring, W. C., 162, 192
Maria Island, 37
Mayhew, Robert, 166, 168, 173
Mayne, R. C., 23
Mekong River, 2
Merilees, Harold, 260
Meyer, Paul, 164
Mica Creek, 197, 199
Ministry of Lands and Forests, 78, 204
Mission, 143
Montana, 201
Moran, 183, 186, 199, 202, 219, 220, 221, 222, 223, 235, 262, 272, 275
  BC Energy Board investigations of, 227, 262
  Cartoon map of, 187
  Dam site, promotion of as a, 186–188, 192–195, 214, 221, 255
Moran Development Corporation, 183, 192–194, 204, 219
Motherwell, J. A., 50, 77, 82
Muir, J. F., 256
Murray, James E., 201
Murrin, W. G., 66, 132

Nahanee, Ed, 225
Nanaimo, 50, 89, 92, 105, 126, 127, 238
Napier, G. P., 31
Nash, Charles, 211, 216
Nass River, 47, 267
National Research Council, 196, 247
Native bands
  Boothroyd, 39, 41, 42, 50
  Cheslatta, *see* Cheslatta T'en
  Cisco, 39
  Coquitlam, 72, 73, 76
  Kwantlen, 76

Native bands (*cont.*)
  North Bend, 39
  Spuzzum, 29, 39, 40
  Yale, 39
Native Brotherhood, 204, 205, 225
Neave, Ferris, 251, 252, 253
Nechako River, 15, 158, 168, 169,
    237, 271, 272, 275
  Development of, 170–173, 175, 176
  Map of, 171
  Power surveys of, 150, 151, 155
  Water rights to, 164, 169–170
Nelles, H. V., 11
Neuberger, Richard, 201–202
New Democratic Party (provincial),
    228
New Westminster, 37, 74, 76, 88
Newell, Dianne, 39
Niagara River, 54
Nicola Valley, 22, 36
Nilsson, Christer, 1, 267
Nisga'a Land Committee, 39
Nlaka'pamux (Thompson Indians),
    21, 22
North Pacific Halibut Commission,
    84, 87, 89, 91, 110
North Thompson River, 223
North Vancouver, 206
Nye, David, 206

Ohamil, 37
Okanagan, 120, 136, 139, 142, 169
Ontario, 125, 127, 239
Ontario Fisheries Research
    Laboratory, 251
Ontario Hydro-Electric Commission,
    125
Ootsa Lake, 171, 175
Oregon, 201, 241
Ottawa, 71, 131, 133, 139, 182

Pacific Biological Station, 105, 110
Pacific Fisherman, 99, 109
Pacific Salmon Convention, 14, 91,
    110, 115, 215, 271, 274

Obstacle of to river development,
    164, 228, 243
  Passage and mandate of, 86, 87
Paget, Arthur, 221
Parkin, Tom, 226, 258
Parsnip River, 226
Paul, James, 29, 37, 42
Paulson, Ed, 161–163
Peace River, 3, 16, 17, 122, 124, 192,
    270, 274, 275
  Damming, politics of, 179–182,
    183, 217, 222, 226, 228–229
Peace River Development
    Corporation, 182
Pechora River, 268
Perry, Harry G. T, 122
Polanyi, Karl, 94
Post-war Rehabilitation Council,
    121–124, 129, 134–135
Potter, Russell, 192, 219, 221, 235
Powell, R. E., 150, 156
Powell River, 257, 258
Pretious, Edward, 101, 102, 259,
    260
Prince George, 121–122, 151, 171,
    212, 223
Prince Rupert Fisherman's
    Cooperative, 225
Pritchard, A. L., 174, 254
Public Utilities Commission (PUC),
    128, 129, 132, 137, 139, 141,
    146
Puget Sound, 44, 46, 47, 48
Purdy, H. L., 184

Quebec, 152, 153, 154
Quesnel, 127, 171, 212
Quesnel dam, 30, 69–72
Quesnel Lake, 75, 82, 99, 103, 104
Quesnel River, 30, 69, 212, 223

Railway Belt, 9
Rawson, D. S., 252
Reid, Tom, 110, 111, 185, 207, 254,
    255, 256, 261

Alcan, criticisms of, 165–166, 168
and the establishment of the IPSFC, 87, 88
Fraser Basin Board, criticisms of, 221, 225
Provincial government, criticisms of, 158–159
Ricker, Bill, 84, 89, 105–113, 116–117
Photograph at Hells Gate, *see* Hells Gate
Rivers Inlet, 47
Rocky Mountain Trench, 179
Royal Geographical Society, 151
Royal, Loyd, 209, 235, 256, 257, 261
Rural electrification, 124–129, 149
Rural Electrification Administration (REA) (U.S.), 125
Rural Electrification Committee, 126, 127, 128–129, 132, 134, 141
Ruskin, 57, 66–67, 68, 130
Ruus, Eugene, 256

Sager, A. H., 185, 208, 209
Saguenay River, 153, 267, 270
Salmon, 5, 20, 21, 140–141, 158, 178, 182, 206–211
and Heritage, 165
Big-year cycles, 19–20
Early theories about migration of, 27–28
Effects of Alcan development on, *see* Alcan
Effects of Hells Gate slides on, *see* Hells Gate
Effects of tributary dams on, 69–71, 83
Fishing, *see* Fishing
Hells Gate and, 19–21, 52
Origin tale of, *see* Coyote
Photo of salmon from Hells Gate investigations, *see* Hells Gate
Scientific study of, *see* fisheries science and Hells Gate
Salmon Arm, 39

Salmon Canners' Operating Committee, 160–161, 169, 212
Saskatchewan River, 267
Scott, Thomas, 49, 95–96
Scuzzy, 24
Scuzzy Rapids, 25, 28, 30, 31, 32, 35
Seton Creek, 141, 202
Seton Lake, 27, 28, 40, 139, 140, 141, 147, 235
Shipshaw, 152
Shrum, Gordon, 209, 220, 226–228, 260, 261, 262
Shuswap Lake, 197
Sinclair, James, 200, 207, 209, 240, 255, 257
Skeena River, 38, 42, 47, 90, 171, 267
Smith, Tim, 112
Social Credit Party (provincial), 122, 179, 194, 219, 227
Sommers, Robert E., 194
Spuzzum, 29, 37, 42, 95
Spuzzum Creek, 29
St. Lawrence River, 2, 267
St. Lawrence Seaway, 186, 213
Stanford, 28, 87, 89, 236
Staples tradition, 10–11
Stave Lake, 57, 62, 64, 66, 68, 76, 77
Stave Lake Power Company, 60, 62, 76
Stave River, 57, 60, 66
Stikine River, 267
Strait of Georgia, 262
Stuart Lake, 172
Summerland, 121
Swinton, Harry, 193
Sword, C. B., 75
Symington, Herbert, 133
System A Plan, 183, 223, 225

Tahtsa Lake, 171, 176
Taiya Project, 156
Talbot, G. B., 114
Taseko Lake, 202
Taylor, Joseph, 12, 249

Teit, James, 23
Tennessee Valley Authority (TVA), 80, 125, 144, 146, 220
Thompson River, 8, 64, 183, 197, 199, 202
Thompson, William F., 84, 86, 87, 91, 93, 100, 102–106, 115
  Approach to Fraser River research of, 89, 90, 91, 99
  Bill Ricker, disagreements with, 106, 107, 108–112, 116
  Biographical background of, 87–88
  IPSFC researchers, recruitment of, 88–89
  Salmon migration difficulties at Hells Gate, analysis of, 93, 98, 100, 102–106
Transnational Environmental History, 11–12
Treaty of Washington (1846), 7
Tweedsmuir, Lord, 170
Tyrrell, Ian, 11

United Fisherman's Cooperative Association, 161
United Fishermen and Allied Workers' Union (UFAWU), 161, 239
  Fraser dams, protests against, 203, 204, 208, 209, 210, 217, 218, 225, 226, 258
University of British Columbia (UBC), 88, 89, 101, 123–124, 209, 231, 238, 239, 254–255, 256, 260
  Fish-Power research at, 219, 225, 226, 240, 242, 246, 248, 250, 251, 258, 259, 260, 265
  Institute of Fisheries of, 232, 238, 239, 244, 246–247, 254–255
University of Saskatchewan, 252
University of Toronto, 105
University of Washington (UW), 84, 87, 88, 89, 101, 108
U.S. Army Corps of Engineers, 80, 81, 101, 220, 236, 241, 273
U.S. Bureau of Fisheries, 108

U.S. Bureau of Reclamation, 80, 81, 156, 220, 273
U.S. Fish and Wildlife Service, 241, 247, 251, 273
U.S. Geological Survey, 78, 79
U.S. Pacific Northwest, 54, 188, 201, 213, 214, 215, 239, 269, 273
U.S. State Department, 164–165, 167

Van Cleve, Richard, 108, 113, 234
Vancouver, 37, 68, 73, 76, 113, 119, 124, 125, 135, 186, 187, 194, 204, 208, 210, 214
  BCER, dealings with, 58
  Early development and population of, 8, 9–10
  Early twentieth century expansion of, 54, 63, 64, 83
  Effects of Fraser River flood on, 142, 144
  Electricity service and supply of, 55, 56, 57, 58, 60, 62, 64, 69, 130, 140, 190, 192
  Electricity shortages in, 66, 133
  Second World War, conditions during, 130, 133
Vancouver Foundation, 239
Vancouver Gas Company, 60
Vancouver Island, 43, 90, 120, 139, 141, 142, 189, 192
Vancouver Power Company, 58
Vancouver Trades and Labour Council, 135
Vanderhoof, 212
Vang, Alfred, 193, 194
Vernon, E. H., 227, 251
Victoria, 69–71, 83, 124, 125, 127, 169, 185, 209, 257

War Production Board, 131
Warren, Harry V., 129, 153, 189, 196, 209, 220, 255, 256, 260, 263
  and the Moran dam, 192–194, 199, 202, 214, 219, 221, 235

BC, promotion of to Alcan, 153
Fraser River, promotion of as a
power source, 184, 186–188, 203
Wartime power development, views
about, 123, 124, 146
Washington, D.C., 152
Washington State, 88, 241, 249
Washington State Department of
Fisheries, 101
Water Act (provincial), 162, 163
Water comptroller, 123, 154, 169,
170, 221
Wenner-Gren, Axel, 179
Wenner-Gren Corporation, 179, 182
West Kootenay Power, 126, 137
West Vancouver, 206

Western Canada Power Company
(WCPC), 62, 64
Weston, Samuel R., 141, 184
Whatashan, 142
White's Creek, 30
Williams Lake, 22, 212
Williston, Ray, 205, 221, 223
Winch, Harold E., 122
World War II, 15, 53, 151, 154, 155,
179, 184, 188, 213, 269, 271
Worster, Donald, 268

Yale, 7, 24, 37, 91, 95
Yangtze River, 2
Yukon River, 2, 156, 267
Yukon Territory, 154, 156

3 5282 00692 9007

Printed in the United States
133463LV00010B/274/A